목 차

Chapter 1. Linear Algebra 3

Chapter 2. Probability Theory 23

Chapter 3. Vector Calculus 51

Chapter 4. Information Theory 69

Chapter 5. Statistics 91

Chapter 6. Inference Overview 123

Chapter 7. Variational Inference 143

Chapter 8. Variational Autoencoder 167

Chapter 9. Diffusion Model 199

Chapter 10. Generative Adversarial Networks 231

에듀컨텐츠 휴피아
Educontents·Huepia

시조배움말

임종기 • 著

초등학교·중학이
CH Educonents Haepa

심층생성모델

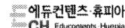

Chapter 1.
Linear Algebra

Vectors Definition

Let

$$u = \begin{bmatrix} u_1 \\ \vdots \\ u_d \end{bmatrix}, \quad v = \begin{bmatrix} v_1 \\ \vdots \\ v_d \end{bmatrix}$$

be two vectors.

Dot Product

The *dot product* of u and v is

$$\langle u, v \rangle = u_1 v_1 + \cdots + u_d v_d$$
$$= \sum_{i=1}^{d} u_i v_i$$

The dot product is an example of an inner product.

Norm of a Vector

The (Euclidean) norm of a vector $u \in \mathbb{R}^n$ is defined by

$$\|u\| := \sqrt{\langle u, u \rangle} = \left(\sum_{j=1}^{n} u_j^2 \right)^{1/2}$$

Example 1

If $u = \begin{bmatrix} 4 \\ 2 \\ -7 \end{bmatrix}$, then $\|u\| = \sqrt{16 + 4 + 49} = \sqrt{69} \approx 8.3$.

Cauchy-Schwarz Inequality

Theorem 2

For any $x, y \in \mathbb{R}^d$, $|\langle x, y \rangle| \le \|x\| \|y\|$, with equality iff x and y are scalar multiples of one another.

Orthogonal Vectors

We say u and v are orthogonal if $u \neq 0 \neq v$, and $\langle u, v \rangle = 0$.

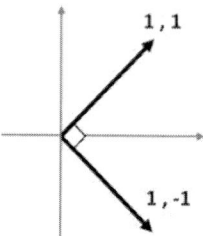

Figure: Two orthogonal vectors.

Vector multiplication

- Inner product: The dot product of u and v can be expressed using matrix multiplication as
$$\langle u, v \rangle = u^T v$$
- Outer product: The product uv^T is a $d \times d$ matrix called the *outer product* of u and v.

Matrix-Vector Multiplication

Let
$$A = \begin{bmatrix} a_{11} & \cdots & a_{1n} \\ \vdots & \ddots & \vdots \\ a_{m1} & \cdots & a_{mn} \end{bmatrix} \in \mathbb{R}^{m \times n}, \quad b = \begin{bmatrix} b_1 \\ \vdots \\ b_n \end{bmatrix} \in \mathbb{R}^n$$

By the definition of matrix multiplication,
$$Ab = \begin{bmatrix} \langle A_{1,:}, b \rangle \\ \vdots \\ \langle A_{m,:}, b \rangle \end{bmatrix} \in \mathbb{R}^m$$

Another Representation

Another very useful way to represent Ab is
$$Ab = \sum_{j=1}^{n} b_j a_j$$

where a_j is the j-th column of A.

Linear Span

A *linear combination* of vectors $a_1, \ldots, a_n \in \mathbb{R}^m$ is any vector of the form

$$\sum_{j=1}^{n} x_j a_j$$

The *span* of $a_1, \ldots, a_n$ is the set of all such linear combinations.

Column Span (colspan(A))

$$\mathrm{span}\{a_1, \ldots, a_n\} = \{y = Ax : x \in \mathbb{R}^n\}$$

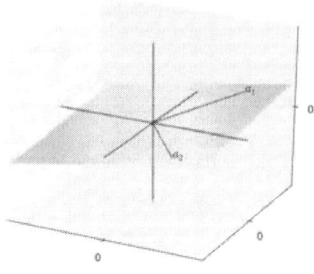

Figure: Linear span of two vectors.

Linear Independence

- Vectors $a_1, \ldots, a_n \in \mathbb{R}^m$ are *linearly independent* iff the following implication holds:

$$\sum_{j=1}^{n} x_j a_j = 0 \implies x_j = 0 \quad \forall j.$$

- The nullspace of matrix A is:

$$N(A) = \{x \in \mathbb{R}^n \mid Ax = 0\}.$$

- Vectors $a_1, \ldots, a_n \in \mathbb{R}^m$ are *linearly independent* iff $N(A) = \{0\}$

Linear Independence (Cont'd.)

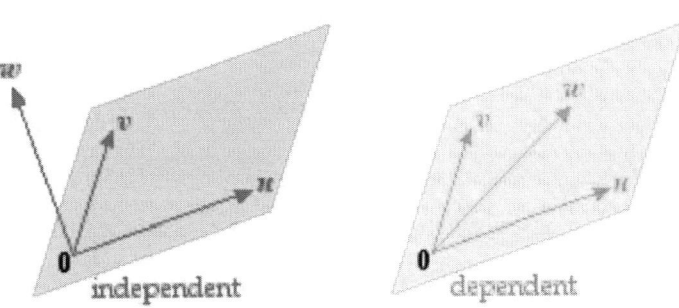

Figure: Left: Linearly independent vectors. Right: Linearly dependent vectors.

Subspaces

A subspace is a set of vectors that is closed under scalar multiplication and vector addition. A set $S \subseteq \mathbb{R}^m$ is a subspace if:

1. For all $\alpha \in \mathbb{R}$ and $u \in S, \quad \alpha u \in S$.
2. For all $u, v \in S, \quad u + v \in S$.

Example 3

For matrix $A \in \mathbb{R}^{m \times n}$, $\mathrm{colspan}(A)$ is a subspace of $\mathbb{R}^m$, and $N(A)$ is a subspace of $\mathbb{R}^n$.

Basis of a Subspace

A *basis* for a subspace $S \subseteq \mathbb{R}^d$ is a vector set B satisfying:

1. B is a minimal spanning set.
2. B is a maximal linearly independent set.
3. $\mathrm{span}(B) = S$ and B is linearly independent set.
4. Every element of S can be written as a unique linear combination of elements of B.

Dimension

The *dimension* of a subspace is the cardinality (number of elements in a set) of a basis.

Rank and Nullity

Given a matrix A, define the *rank* of A as:

$$\mathrm{rank}(A) = \dim(\mathrm{colspan}(A))$$

and the *nullity* as:

$$\mathrm{nullity}(A) = \dim(N(A)).$$

> **Theorem 4 (Rank-Nullity Theorem)**
> If $A \in \mathbb{R}^{m \times n}$, then
> $$\mathrm{rank}(A) + \mathrm{nullity}(A) = n.$$

Example

Example 5

Consider a matrix $A = \begin{bmatrix} 1 & 2 \\ 3 & 4 \\ 5 & 6 \end{bmatrix}$.

- ① Rank: $\text{rank}(A) = 2$, since it has two linearly independent columns.
- ② Nullity: $\text{nullity}(A) = 0$, since the nullspace contains only the zero vector.

According to the Rank-Nullity theorem:
$\text{rank}(A) + \text{nullity}(A) = 2 + 0 = 2$, which is the number of columns.

Inverses

A square matrix $A \in \mathbb{R}^{d \times d}$ is invertible when there exists another square matrix B such that $AB = BA = I$, the identity matrix. The matrix A is invertible iff:

- ① $\text{rank}(A) = d$
- ② The columns of A are linearly independent
- ③ $\text{nullity}(A) = 0$
- ④ $Ax = 0$ implies $x = 0$
- ⑤ $\det(A) \neq 0$
- ⑥ The equation $Ax = b$ has a unique solution for all $b \in \mathbb{R}^d$.

Orthogonal and Orthonormal Sets

An orthogonal set of vectors is denoted as $\{u_1, \ldots, u_m\}$ where the inner product of any two distinct vectors is zero:

$$\langle u_i, u_j \rangle = 0 \quad \text{for all } i \neq j$$

If $\|u_i\| = 1$ for all i, the set is called orthonormal.

Example 6

$$\begin{bmatrix} 1 \\ 0 \\ 0 \end{bmatrix}, \begin{bmatrix} 0 \\ 1 \\ 0 \end{bmatrix}, \begin{bmatrix} 0 \\ 0 \\ 1 \end{bmatrix}$$

are orthonormal vectors.

Orthogonal Sets and Linearly Independence

Theorem 7
Every orthogonal set is linearly independent.

Proof.
Suppose $u_1, \ldots, u_n$ are orthogonal and $c_1 u_1 + \cdots + c_n u_n = 0$ for some scalars $c_1, \ldots, c_n \in \mathbb{R}$. Then for each i

$$\begin{aligned} 0 &= \langle 0, u_i \rangle \\ &= \langle c_1 u_1 + \cdots + c_n u_n, u_i \rangle \\ &= c_i \|u_i\|^2 \end{aligned}$$

which implies $c_i = 0$ since $\|u_i\| > 0$.

Orthogonal Basis

If $\{u_1, \ldots, u_m\}$ is an orthonormal set and a basis of a subspace S, it is referred to as an *orthonormal basis* (ONB) of that subspace.

Orthogonal Matrix

An *orthogonal matrix* U satisfies:
$$UU^T = U^TU = I$$

Example 8

The rotation matrix
$$\begin{bmatrix} \cos\theta & \sin\theta \\ -\sin\theta & \cos\theta \end{bmatrix}$$
is an orthogonal matrix.

Eigenvalues and Eigenvectors

Let $A \in \mathbb{R}^{d \times d}$. If
$$Au = \lambda u$$
for some $\lambda \in \mathbb{R}$ and $u \in \mathbb{R}^d$, then λ is an eigenvalue of A and u is a corresponding eigenvector.

Spectral Theorem

Theorem 9

If A is symmetric, then
$$A = U \Lambda U^T$$
where U is an orthogonal matrix and Λ is a diagonal matrix.

Spectral Theorem (Cont'd.)

Multiplying $A = U\Lambda U^T$ on the right by U, we have

$$AU = U\Lambda.$$

Let

$$U = \begin{bmatrix} | & & | \\ u_1 & \cdots & u_d \\ | & & | \end{bmatrix}$$

and

$$\Lambda = \begin{bmatrix} \lambda_1 & \cdots & 0 \\ \vdots & \ddots & \vdots \\ 0 & \cdots & \lambda_d \end{bmatrix}$$

If we look at the matrix equation $AU = U\Lambda$ one column at a time, we have

$$Au_i = \lambda_i u_i, \quad i = 1, \ldots, n.$$

$\to \lambda_i$ are eigenvalues of A, and u_i are corresponding eigenvectors

Spectral Theorem (Cont'd.)

An important identity is

$$U\Lambda U^T = \sum_{i=1}^{d} \lambda_i u_i u_i^T$$

This identity expresses A as a sum of rank 1 outer products, known as the spectral or eigenvalue decomposition of A.

Determinant and Trace

- The determinant of a matrix is the product of its eigenvalues.
- The trace of a matrix is the sum of its eigenvalues.

Positive (Semi-)Definite Matrices

Let $A \in \mathbb{R}^{d \times d}$ be a square matrix. We say

- A is *positive semi-definite* if $x^T A x \geq 0$ for all $x \in \mathbb{R}^d$
- A is *positive definite* if $x^T A x > 0$ for all $x \neq 0$.

If A is PD, it is also PSD.

Positive (Semi-)Definite Matrices

PD and PSD matrices arise frequently in machine learning. For example,

- Gram matrices:
$$\begin{bmatrix} \langle x_1, x_1 \rangle & \cdots & \langle x_1, x_n \rangle \\ \vdots & \ddots & \vdots \\ \langle x_n, x_1 \rangle & \cdots & \langle x_n, x_n \rangle \end{bmatrix}$$
- Covariance matrices
- Kernel matrices
- Hessian matrices of convex functions

Characterizing PSD and PD matrices

- A is PSD iff $\lambda_i \geq 0 \quad \forall i$.
- A is PD iff $\lambda_i > 0 \quad \forall i$.

Orthogonal Complements

Let $S \subseteq \mathbb{R}^d$ be a subspace. The set
$$S^\perp := \{v \mid \langle u, v \rangle = 0 \text{ for all } u \in S\}$$
is called the *orthogonal complement* of S.

Orthogonal Complements

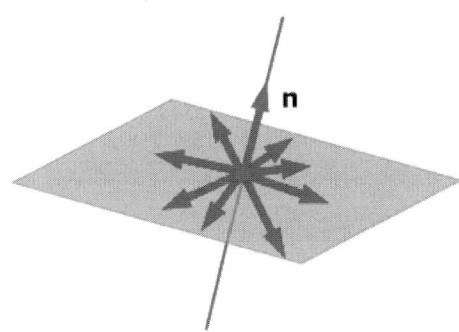

Figure: Illustration of orthogonal complement.

Projection Mapping

Let $\langle A \rangle$ be the column span of A.

The *projection* onto subspace $S = \langle A \rangle$ is the mapping Π_A such that:

$$\Pi_A x = A \widehat{\theta}$$

where

$$\widehat{\theta} = \arg\min_{\theta \in \mathbb{R}^k} \|x - A\theta\|^2.$$

We find that

$$\widehat{\theta} = \left(A^T A\right)^{-1} A^T x$$

And therefore,

$$\Pi_A x = A \left(A^T A\right)^{-1} A^T x$$

Linear Span and Projection

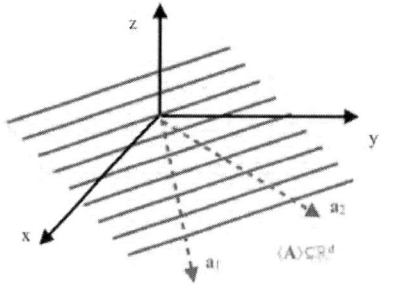
(a) Column span of A where $d = 3, k = 2$.

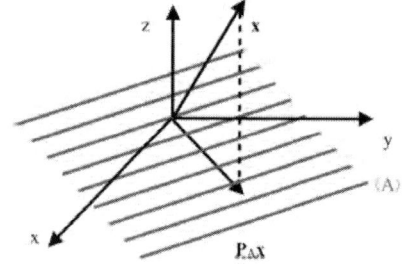
(b) Projection of x onto $\langle A \rangle$.

Projection Matrix

- The *projection matrix* for A is:

$$\Pi_A := A\left(A^T A\right)^{-1} A^T$$

- Projection matrices are symmetric and positive-semidefinite.
- If the columns are orthonormal, i.e.,

$$\langle a_i, a_j \rangle = \begin{cases} 1 \text{ if } i = j \\ 0 \text{ if } i \neq j \end{cases}$$

then
$$\Pi_A = AA^T \quad (d \times d)$$

- Projection matrices are *idempotent*:

$$\Pi_A^2 = \Pi_A$$

Orthogonality Principle

The *orthogonality principle* states that for all x, $x - \Pi_A x$ is orthogonal to every element of $\langle A \rangle$.

$$x - \Pi_A x \in \langle A \rangle^\perp$$

Projection Theorem

The *projection theorem* states that every vector can be uniquely written as:

$$x = u + v, \quad \text{where} \quad u \in S, \quad v \in S^\perp$$

강 의 노 트

1.

2.

3.

Chapter 2.
Probability Theory

Introduction

Probability theory is the study of uncertainty. In this course, we rely on probability theory for deriving machine learning algorithms.

Sample Space

- *Sample space* Ω: The set of all outcomes of a random experiment.
- Each outcome $\omega \in \Omega$ is a complete description of the state of the world at the end of the experiment.

Set of Events

- *Set of events (event space)* $\mathcal{F}$: Elements $A \in \mathcal{F}$ are subsets of Ω.
- $\mathcal{F}$ must satisfy:
  1. $\emptyset \in \mathcal{F}$
  2. $A \in \mathcal{F} \implies \Omega \setminus A \in \mathcal{F}$
  3. $A_1, A_2, \ldots \in \mathcal{F} \implies \bigcup_i A_i \in \mathcal{F}$

Probability Measure

Probability measure: $P : \mathcal{F} \to \mathbb{R}$ satisfying the Axioms of Probability:

1. $P(A) \geq 0$ for all $A \in \mathcal{F}$
2. $P(\Omega) = 1$
3. If $A_1, A_2, \ldots$ are disjoint, then

$$P\left(\bigcup_i A_i\right) = \sum_i P(A_i)$$

Example: Tossing a Die

- Sample Space: $\Omega = \{1, 2, 3, 4, 5, 6\}$
- Event Space example: $\mathcal{F} = \{\emptyset, \Omega\}$
  Probability Measure: $P(\emptyset) = 0, P(\Omega) = 1$
- another Event Space example: $\mathcal{F} =$ the power set of Ω
  Probability Measure: $P(A) = \frac{|A|}{6}$

Additional Properties

- $A \subseteq B \implies P(A) \leq P(B)$
- $P(A \cap B) \leq \min(P(A), P(B))$
- $P(A \cup B) \leq P(A) + P(B)$ (Union Bound)
- $P(\Omega \backslash A) = 1 - P(A)$

Law of Total Probability

If $A_1, \ldots, A_k$ are disjoint events such that $\bigcup_{i=1}^{k} A_i = \Omega$,

$$\sum_{i=1}^{k} P(A_k) = 1$$

Conditional Probability

Let B be an event with $P(B) \neq 0$. The conditional probability $P(A|B)$ is defined as:
$$P(A|B) = \frac{P(A \cap B)}{P(B)}$$

Independence

Two events A and B are independent if:

$$P(A \cap B) = P(A)P(B)$$

Equivalently, $P(A|B) = P(A)$, i.e., observing B does not affect the probability of A.

Random Variables

Consider an experiment where we flip 10 coins and want to know the number of coins that show heads. The sample space Ω consists of 10-length sequences of heads and tails.

$$w_0 = \langle H, H, T, H, T, H, H, T, T, T \rangle \in \Omega$$

We are usually interested in real-valued functions of these outcomes, known as *random variables*.

Formal Definition

A random variable X is defined as a function:

$$X : \Omega \longrightarrow \mathbb{R}$$

We denote random variables using uppercase letters $X(\omega)$ or simply X, and the value that a random variable may take on using lowercase letters x.

Example 1: Discrete Random Variable

Suppose $X(\omega)$ represents the number of heads in ω. Since there are only 10 coins, $X(\omega)$ can only take finite values, making it a *discrete random variable*.

$$P(X = k) := P(\{\omega : X(\omega) = k\})$$

Example 2: Continuous Random Variable

Suppose $X(\omega)$ represents the time for a radioactive particle to decay. $X(\omega)$ can take an infinite number of values, making it a *continuous random variable*.

$$P(a \leq X \leq b) := P(\{\omega : a \leq X(\omega) \leq b\})$$

Cumulative Distribution Functions (CDF)

A CDF $F_X(x)$ is defined as:

$$F_X(x) \triangleq P(X \leq x)$$

CDF example

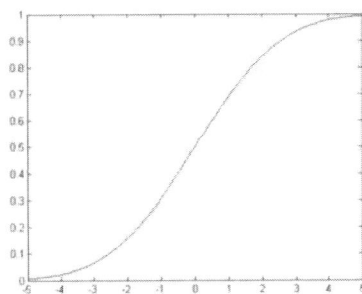

Figure: A Cumulative Distribution Function (CDF)

Properties of CDF

- $0 \leq F_X(x) \leq 1$
- $\lim_{x \to -\infty} F_X(x) = 0$
- $\lim_{x \to \infty} F_X(x) = 1$
- $x \leq y \implies F_X(x) \leq F_X(y)$

Probability Mass Functions (PMF)

A Probability Mass Function (PMF) is a function:

$$p_X(x) \triangleq P(X = x).$$

It is used when X is a discrete random variable.

Notation for Discrete Random Variables

For a discrete random variable, we use the notation $\mathrm{Val}(X)$ for the set of possible values.

Example 10

If $X(\omega)$ is the number of heads in 10 coin tosses, then $\mathrm{Val}(X) = \{0, 1, 2, \ldots, 10\}$.

Properties of PMF

Property 1
- $0 \leq p_X(x) \leq 1$
- $\sum_{x \in \mathrm{Val}(X)} p_X(x) = 1$
- $\sum_{x \in A} p_X(x) = P(X \in A)$

Probability Density Functions (PDF)

For continuous random variables, we define the Probability Density Function (PDF) as:

$$f_X(x) \triangleq \frac{dF_X(x)}{dx}.$$

Properties of PDF

Property 2
- $f_X(x) \geq 0$
- $\int_{-\infty}^{\infty} f_X(x) dx = 1$
- $\int_{x \in A} f_X(x)\, dx = P(X \in A)$

Expectation

Suppose that X is a discrete random variable with PMF $p_X(x)$ and $g : \mathbb{R} \longrightarrow \mathbb{R}$ is an arbitrary function.

The *expectation* or *expected value* of $g(X)$ is defined as:

$$E[g(X)] \triangleq \sum_{x \in \text{Val}(X)} g(x) p_X(x).$$

For continuous random variables:

$$E[g(X)] \triangleq \int_{-\infty}^{\infty} g(x) f_X(x)\, dx$$

Properties of Expectation

Property 3
- $E[a] = a$ for any constant $a \in \mathbb{R}$.
- $E[af(X)] = aE[f(X)]$ for any constant $a \in \mathbb{R}$.
- $E[f(X) + g(X)] = E[f(X)] + E[g(X)]$
- $E[1\{X = k\}] = P(X = k)$

Variance

Definition

The *variance* of a random variable X is a measure of how concentrated the distribution of X is around its mean:

$$\text{Var}[X] \triangleq E\left[(X - E[X])^2\right]$$

Alternative Expression for Variance

$$E\left[(X - E[X])^2\right] = E\left[X^2 - 2E[X]X + E[X]^2\right]$$
$$= E[X^2] - E[X]^2$$

Properties of Variance

- $\text{Var}[a] = 0$ for any constant $a \in \mathbb{R}$.
- $\text{Var}[af(X)] = a^2\text{Var}[f(X)]$ for any constant $a \in \mathbb{R}$.

Joint and Marginal PMFs for Discrete Random Variables

$$p_{XY}(x,y) = P(X=x, Y=y)$$
$$p_X(x) = \sum_y p_{XY}(x,y)$$

Process of forming the marginal distribution is called *marginalization*.

Joint and Marginal PDFs for Continuous Random Variables

$$f_{XY}(x,y) = \frac{\partial^2 F_{XY}(x,y)}{\partial x \partial y}$$
$$f_X(x) = \int_{-\infty}^{\infty} f_{XY}(x,y) dy$$

Conditional Distributions

Discrete Case:

$$p_{Y|X}(y|x) = \frac{p_{XY}(x,y)}{p_X(x)}$$

Continuous Case:

$$f_{Y|X}(y|x) = \frac{f_{XY}(x,y)}{f_X(x)}$$

Bayes's Rule

Discrete Case:

$$P_{Y|X}(y|x) = \frac{P_{XY}(x,y)}{P_X(x)} = \frac{P_{X|Y}(x|y)P_Y(y)}{\sum_{y'} P_{X|Y}(x|y')P_Y(y')}$$

Continuous Case:

$$f_{Y|X}(y|x) = \frac{f_{XY}(x,y)}{f_X(x)} = \frac{f_{X|Y}(x|y)f_Y(y)}{\int_{-\infty}^{\infty} f_{X|Y}(x|y')f_Y(y')dy'}$$

Independence of Random Variables

Two random variables X and Y are *independent* if:

- $F_{XY}(x,y) = F_X(x)F_Y(y)$
- For discrete RVs, $p_{XY}(x,y) = p_X(x)p_Y(y)$
- For continuous RVs, $f_{XY}(x,y) = f_X(x)f_Y(y)$

Lemma on Independence

If X and Y are independent then for any subsets $A, B \subseteq \mathbb{R}$,

$$P(X \in A, Y \in B) = P(X \in A)P(Y \in B)$$

Expectation and Covariance

Suppose that we have two discrete random variables X, Y. Then, the expected value of $g(X, Y)$ is:

$$E[g(X,Y)] \triangleq \sum_{x \in \text{Val}(X)} \sum_{y \in \text{Val}(Y)} g(x,y) p_{XY}(x,y)$$

For continuous X, Y, the analogous expression is:

$$E[g(X,Y)] = \int_{-\infty}^{\infty} \int_{-\infty}^{\infty} g(x,y) f_{XY}(x,y) dx dy$$

Covariance

The covariance of X and Y is defined as:

$$\text{Cov}[X,Y] \triangleq E[(X - E[X])(Y - E[Y])]$$

Using linearity properties, covariance can be rewritten as:

$$\text{Cov}[X,Y] = E[XY] - E[X]E[Y]$$

Note: $E[X]$ and $E[Y]$ are constants.

Important Properties

- Linearity of expectation:
  $E[f(X,Y) + g(X,Y)] = E[f(X,Y)] + E[g(X,Y)]$
- $\text{Var}[X+Y] = \text{Var}[X] + \text{Var}[Y] + 2\text{Cov}[X,Y]$
- If X and Y are independent, $\text{Cov}[X,Y] = 0$
- If X and Y are independent, $E[f(X)g(Y)] = E[f(X)]E[g(Y)]$

Multiple Random Variables

The ideas can be generalized to n continuous random variables $X_1, X_2, \ldots, X_n$.

- Joint Distribution Function: $F_{X_1, X_2, \ldots, X_n}$
- Joint PDF: $f_{X_1, X_2, \ldots, X_n}$
- Marginal PDF: f_{X_1}
- Conditional PDF: $f_{X_1 | X_2, \ldots, X_n}$

Calculating Probabilities

To calculate the probability of an event $A \subseteq \mathbb{R}^n$:

$$P((x_1, x_2, \ldots, x_n) \in A) = \int_{(x_1, x_2, \ldots, x_n) \in A} f_{X_1, X_2, \ldots, X_n} dx_1 dx_2 \ldots dx_n$$

Chain Rule

Chain rule for multiple random variables:

$$f(x_1, x_2, \ldots, x_n) = f(x_1) \prod_{i=2}^{n} f(x_i \mid x_1, \ldots, x_{i-1})$$

Why Gaussian Distributions are Important

- Common when modeling "noise"
- Convenient for analytical manipulations
- Many integrals involving Gaussian distributions have closed-form solutions

Covariance Matrix

For a random vector X, the covariance matrix Σ is defined as:

$$\Sigma = E\left[(X - E[X])(X - E[X])^T\right]$$

Properties:

- $\Sigma \succeq 0$ (Positive semidefinite)
- $\Sigma = \Sigma^T$ (Symmetric)

The Multivariate Gaussian Distribution

A random vector X has a multivariate normal distribution if:

$$\begin{aligned}
&f_{X_1,X_2,\ldots,X_n}(x_1, x_2, \ldots, x_n; \mu, \Sigma) \\
&= \frac{1}{(2\pi)^{n/2}|\Sigma|^{1/2}} \exp\left(-\frac{1}{2}(x-\mu)^T \Sigma^{-1}(x-\mu)\right)
\end{aligned}$$

$$\iff X \sim \mathcal{N}(\mu, \Sigma)$$

Expectation

For an arbitrary function $g : \mathbb{R}^n \to \mathbb{R}$, the expected value is defined as:

$$E[g(X)] = \int_{\mathbb{R}^n} g(x_1, x_2, \ldots, x_n) f_{X_1, X_2, \ldots, X_n}(x_1, x_2, \ldots x_n) \, dx_1 dx_2 \ldots dx_n,$$

Vector-valued Expectation

If g is a function from $\mathbb{R}^n$ to $\mathbb{R}^m$, the expected value is:

$$E[g(X)] = \begin{bmatrix} E[g_1(X)] \\ E[g_2(X)] \\ \vdots \\ E[g_m(X)] \end{bmatrix}$$

Independence

Random variables $X_1, \ldots, X_n$ are independent if:

$$f(x_1, \ldots, x_n) = f(x_1)f(x_2) \cdots f(x_n)$$

Random Vectors

Suppose that we have n random variables. When working with all these random variables together, we often find it convenient to put them in a vector X:

$$X = \begin{bmatrix} X_1 & X_2 & \ldots & X_n \end{bmatrix}^T$$

This is called a *random vector*.

Chapter 3.
Vector Calculus

Introduction

We briefly revisit differentiation of a univariate function, which may be familiar from high school mathematics. We start with the difference quotient of a univariate function $y = f(x), x, y \in \mathbb{R}$, which we will use to define derivatives.

Difference Quotient

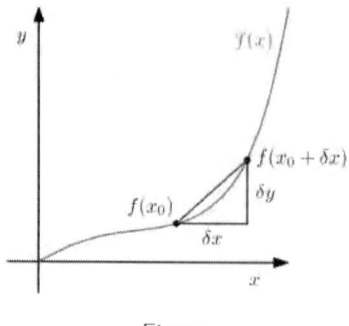

Figure

Difference Quotient (Cont'd.)

Definition

The difference quotient is defined as:
$$\frac{\delta y}{\delta x} := \frac{f(x + \delta x) - f(x)}{\delta x}$$

This computes the slope of the secant line through two points on the graph of f. The difference quotient can be considered as the average slope of f between x and $x + \delta x$ for a linear function f.

Derivative

Definition

The derivative of f at x is:
$$\frac{\mathrm{d}f}{\mathrm{d}x} := \lim_{h \to 0} \frac{f(x + h) - f(x)}{h}$$

This points in the direction of steepest ascent of f.

Example: Derivative of a Polynomial

Consider $f(x) = x^n$, where $n \in \mathbb{N}$. Using the definition of the derivative, we get:

$$\frac{\mathrm{d}f}{\mathrm{d}x} = \lim_{h \to 0} \frac{(x+h)^n - x^n}{h}$$

$$= \lim_{h \to 0} \frac{\sum_{i=0}^{n} \binom{n}{i} x^{n-i} h^i - x^n}{h}$$

Simplifying, we obtain:

$$\frac{\mathrm{d}f}{\mathrm{d}x} = \lim_{h \to 0} \sum_{i=1}^{n} \binom{n}{i} x^{n-i} h^{i-1}$$

$$= \binom{n}{1} x^{n-1} = n x^{n-1}$$

Taylor Series: Introduction

The Taylor series is a representation of a function f as an infinite sum of terms. These terms are determined using derivatives of f evaluated at x_0.

Taylor Polynomial

Definition 13 (Taylor Polynomial)

The Taylor polynomial of degree n of $f : \mathbb{R} \to \mathbb{R}$ at x_0 is defined as

$$T_n(x) := \sum_{k=0}^{n} \frac{f^{(k)}(x_0)}{k!} (x - x_0)^k,$$

where $f^{(k)}(x_0)$ is the k-th derivative of f at x_0 and $\frac{f^{(k)}(x_0)}{k!}$ are the coefficients of the polynomial.

Taylor Series

Definition 14 (Taylor Series)

For a smooth function $f \in \mathcal{C}^\infty, f : \mathbb{R} \to \mathbb{R}$, the Taylor series of f at x_0 is defined as

$$T_\infty(x) = \sum_{k=0}^{\infty} \frac{f^{(k)}(x_0)}{k!} (x - x_0)^k.$$

If $f(x) = T_\infty(x)$, then f is called analytic.

Example: $f(x) = x^4$

We consider the polynomial
$$f(x) = x^4$$
and seek the Taylor polynomial T_6 evaluated at $x_0 = 1$. We start by computing the coefficients $f^{(k)}(1)$ for $k = 0, \ldots, 6$:

$$f(1) = 1$$
$$f'(1) = 4$$
$$f''(1) = 12$$
$$f^{(3)}(1) = 24$$
$$f^{(4)}(1) = 24$$
$$f^{(5)}(1) = 0$$
$$f^{(6)}(1) = 0$$

Therefore, the desired Taylor polynomial is
$$T_6(x) = 1 + 4(x-1) + 6(x-1)^2$$
$$+ 4(x-1)^3 + (x-1)^4 + 0$$

Example: $f(x) = x^4$ (Cont'd)

Multiplying out and re-arranging yields
$$T_6(x) = (1 - 4 + 6 - 4 + 1) + x(4 - 12 + 12 - 4)$$
$$+ x^2(6 - 12 + 6) + x^3(4 - 4) + x^4$$
$$= x^4 = f(x)$$

i.e., we obtain an exact representation of the original function.

Differentiation Rules

In the following, we briefly state basic differentiation rules. We denote the derivative of f by f'.

- Product rule:
$$(f(x)g(x))' = f'(x)g(x) + f(x)g'(x)$$
- Quotient rule:
$$\left(\frac{f(x)}{g(x)}\right)' = \frac{f'(x)g(x) - f(x)g'(x)}{(g(x))^2}$$
- Sum rule:
$$(f(x) + g(x))' = f'(x) + g'(x)$$
- Chain rule:
$$(g(f(x)))' = (g \circ f)'(x) = g'(f(x))f'(x) \qquad (1)$$

Here, $g \circ f$ denotes function composition $x \mapsto f(x) \mapsto g(f(x))$.

Example: Chain Rule for $h(x) = (2x+1)^4$

Let us compute the derivative of the function $h(x) = (2x+1)^4$ using the chain rule.

With
$$h(x) = (2x+1)^4 = g(f(x)), \quad f(x) = 2x+1, \quad g(f) = f^4,$$
we obtain the derivatives of f and g as
$$f'(x) = 2, \quad g'(f) = 4f^3.$$

The derivative of h is given as
$$h'(x) = g'(f)f'(x) = (4f^3) \cdot 2 = 8(2x+1)^3,$$
where we used the chain rule and substituted the definition of f in $g'(f)$.

Introduction

Previously discussed differentiation applies to functions f of a scalar variable $x \in \mathbb{R}$.

Now we consider the case where f depends on one or more variables $\boldsymbol{x} \in \mathbb{R}^n$, e.g., $f(\boldsymbol{x}) = f(x_1, x_2)$.

The generalization of the derivative to functions of several variables is the gradient.

Gradient Computation

We find the gradient of the function f with respect to $\boldsymbol{x}$ by varying one variable at a time and keeping the others constant. The gradient is the collection of these partial derivatives.

Partial Derivative

Definition 15 (Partial Derivative)

For a function $f : \mathbb{R}^n \to \mathbb{R}, \boldsymbol{x} \mapsto f(\boldsymbol{x}), \boldsymbol{x} \in \mathbb{R}^n$ of n variables $x_1, \ldots, x_n$, we define the partial derivatives as

$$\frac{\partial f}{\partial x_1} = \lim_{h \to 0} \frac{f(x_1 + h, x_2, \ldots, x_n) - f(\boldsymbol{x})}{h},$$

$$\vdots,$$

$$\frac{\partial f}{\partial x_n} = \lim_{h \to 0} \frac{f(x_1, \ldots, x_{n-1}, x_n + h) - f(\boldsymbol{x})}{h}.$$

Partial Derivative (Cont'd)

We collect these partial derivatives in the row vector

$$\nabla_{\boldsymbol{x}} f = \operatorname{grad} f = \frac{\mathrm{d} f}{\mathrm{d} \boldsymbol{x}} = \left[\frac{\partial f(\boldsymbol{x})}{\partial x_1}, \frac{\partial f(\boldsymbol{x})}{\partial x_2}, \ldots, \frac{\partial f(\boldsymbol{x})}{\partial x_n} \right] \in \mathbb{R}^{1 \times n}, \quad (2)$$

where n is the number of variables and 1 is the dimension of the image/range/codomain of f.

Here, we defined the column vector $\boldsymbol{x} = [x_1, \ldots, x_n]^\top \in \mathbb{R}^n$. The row vector in equation (2) is called the gradient of f or the Jacobian.

Example

For $f(x,y) = (x + 2y^3)^2$, the partial derivatives are:

$$\frac{\partial f(x,y)}{\partial x} = 2(x + 2y^3)\frac{\partial}{\partial x}(x + 2y^3)$$
$$= 2(x + 2y^3),$$

$$\frac{\partial f(x,y)}{\partial y} = 2(x + 2y^3)\frac{\partial}{\partial y}(x + 2y^3)$$
$$= 12(x + 2y^3)y^2,$$

where we used the chain rule Eq. (1).

Remark: Gradient as a Row Vector

It is common to define the gradient vector as a column vector, but we define it as a row vector for two reasons:

- To generalize to vector-valued functions $f : \mathbb{R}^n \to \mathbb{R}^m$ (the gradient becomes a matrix).
- To apply the multi-variate chain rule without worrying about dimensions.

Example

For $f(x_1, x_2) = x_1^2 x_2 + x_1 x_2^3$, the partial derivatives are:

$$\frac{\partial f(x_1, x_2)}{\partial x_1} = 2x_1 x_2 + x_2^3,$$

$$\frac{\partial f(x_1, x_2)}{\partial x_2} = x_1^2 + 3x_1 x_2^2,$$

The gradient is then:

$$\frac{\mathrm{d}f}{\mathrm{d}\boldsymbol{x}} = \left[\frac{\partial f(x_1, x_2)}{\partial x_1}, \frac{\partial f(x_1, x_2)}{\partial x_2}\right] = \left[2x_1 x_2 + x_2^3, x_1^2 + 3x_1 x_2^2\right] \in \mathbb{R}^{1 \times 2}.$$

Basic Rules of Partial Differentiation

In the multivariate case, where $x \in \mathbb{R}^n$, the basic differentiation rules still apply. However, we need to pay attention: Our gradients now involve vectors and matrices, and matrix multiplication is not commutative, i.e., the order matters.

General Rules

Basic Rules of Partial Differentiation

Here are the general product rule, sum rule, and chain rule:

- Product rule:
$$\frac{\partial}{\partial \boldsymbol{x}}(f(\boldsymbol{x})g(\boldsymbol{x})) = \frac{\partial f}{\partial \boldsymbol{x}}g(\boldsymbol{x}) + f(\boldsymbol{x})\frac{\partial g}{\partial \boldsymbol{x}}$$

- Sum rule:
$$\frac{\partial}{\partial \boldsymbol{x}}(f(\boldsymbol{x}) + g(\boldsymbol{x})) = \frac{\partial f}{\partial \boldsymbol{x}} + \frac{\partial g}{\partial \boldsymbol{x}}$$

- Chain rule:
$$\frac{\partial}{\partial \boldsymbol{x}}(g \circ f)(\boldsymbol{x}) = \frac{\partial g}{\partial f}\frac{\partial f}{\partial \boldsymbol{x}}$$

Chain Rule

Basic Rules of Partial Differentiation   Chain Rule

Consider a function $f : \mathbb{R}^2 \to \mathbb{R}$ of two variables x_1, x_2. Furthermore, $x_1(t)$ and $x_2(t)$ are themselves functions of t.

$$\frac{\mathrm{d}f}{\mathrm{d}t} = \frac{\partial f}{\partial x_1}\frac{\partial x_1}{\partial t} + \frac{\partial f}{\partial x_2}\frac{\partial x_2}{\partial t},$$

where d denotes the gradient and ∂ partial derivatives.

Example

Consider $f(x_1, x_2) = x_1^2 + 2x_2$, where $x_1 = \sin t$ and $x_2 = \cos t$,

$$\frac{df}{dt} = 2\sin t(\cos t - 1)$$

is the corresponding derivative of f with respect to t.

Chain Rule for Functions of Two Variables

If $f(x_1, x_2)$ is a function of x_1 and x_2, where $x_1(s,t)$ and $x_2(s,t)$ are themselves functions of two variables s and t,

$$\frac{\partial f}{\partial s} = \frac{\partial f}{\partial x_1}\frac{\partial x_1}{\partial s} + \frac{\partial f}{\partial x_2}\frac{\partial x_2}{\partial s},$$

$$\frac{\partial f}{\partial t} = \frac{\partial f}{\partial x_1}\frac{\partial x_1}{\partial t} + \frac{\partial f}{\partial x_2}\frac{\partial x_2}{\partial t},$$

The gradient is obtained by the matrix multiplication:

$$\frac{df}{d(s,t)} = \frac{\partial f}{\partial \boldsymbol{x}}\frac{\partial \boldsymbol{x}}{\partial (s,t)} = \underbrace{\begin{bmatrix} \frac{\partial f}{\partial x_1} & \frac{\partial f}{\partial x_2} \end{bmatrix}}_{=\frac{\partial f}{\partial \boldsymbol{x}}} \underbrace{\begin{bmatrix} \frac{\partial x_1}{\partial s} & \frac{\partial x_1}{\partial t} \\ \frac{\partial x_2}{\partial s} & \frac{\partial x_2}{\partial t} \end{bmatrix}}_{=\frac{\partial \boldsymbol{x}}{\partial (s,t)}}.$$

Introduction to Vector-Valued Functions

We've discussed partial derivatives and gradients for functions $f : \mathbb{R}^n \to \mathbb{R}$. Now, let's generalize to vector-valued functions $f : \mathbb{R}^n \to \mathbb{R}^m$ where $n \geq 1$ and $m > 1$.

Vector of Function Values

For a function $\boldsymbol{f} : \mathbb{R}^n \to \mathbb{R}^m$ and vector $\boldsymbol{x}$,

$$\boldsymbol{f}(\boldsymbol{x}) = \begin{bmatrix} f_1(\boldsymbol{x}) \\ \vdots \\ f_m(\boldsymbol{x}) \end{bmatrix} \in \mathbb{R}^m$$

This allows us to view $\boldsymbol{f}$ as a vector of functions $[f_1, \ldots, f_m]^\top$ ($f_i : \mathbb{R}^n \to \mathbb{R}$).

Partial Derivative of Vector-Valued Functions

The partial derivative of $\boldsymbol{f}: \mathbb{R}^n \to \mathbb{R}^m$ with respect to x_i is:

$$\frac{\partial \boldsymbol{f}}{\partial x_i} = \begin{bmatrix} \frac{\partial f_1}{\partial x_i} \\ \vdots \\ \frac{\partial f_m}{\partial x_i} \end{bmatrix} = \begin{bmatrix} \lim_{h \to 0} \frac{f_1(x_1,\ldots,x_{i-1},x_i+h,x_{i+1},\ldots x_n)-f_1(\boldsymbol{x})}{h} \\ \vdots \\ \lim_{h \to 0} \frac{f_m(x_1,\ldots,x_{i-1},x_i+h,x_{i+1},\ldots x_n)-f_m(\boldsymbol{x})}{h} \end{bmatrix} \in \mathbb{R}^m \quad (3)$$

Gradient of Vector-Valued Functions

The gradient of $\boldsymbol{f}$ is obtained by collecting these partial derivatives:

$$\frac{\mathrm{d}\boldsymbol{f}(\boldsymbol{x})}{\mathrm{d}\boldsymbol{x}} = \begin{bmatrix} \frac{\partial \boldsymbol{f}(\boldsymbol{x})}{\partial x_1} & \cdots & \frac{\partial \boldsymbol{f}(\boldsymbol{x})}{\partial x_n} \end{bmatrix}$$

$$= \begin{bmatrix} \frac{\partial f_1(\boldsymbol{x})}{\partial x_1} & \cdots & \frac{\partial f_1(\boldsymbol{x})}{\partial x_n} \\ \vdots & & \vdots \\ \frac{\partial f_m(\boldsymbol{x})}{\partial x_1} & \cdots & \frac{\partial f_m(\boldsymbol{x})}{\partial x_n} \end{bmatrix} \in \mathbb{R}^{m \times n}.$$

Jacobian

Definition 16 (Jacobian)
The Jacobian J is an $m \times n$ matrix of all first-order partial derivatives:

$$J = \frac{\mathrm{d}f(x)}{\mathrm{d}x} = \begin{bmatrix} \frac{\partial f_1(x)}{\partial x_1} & \cdots & \frac{\partial f_1(x)}{\partial x_n} \\ \vdots & & \vdots \\ \frac{\partial f_m(x)}{\partial x_1} & \cdots & \frac{\partial f_m(x)}{\partial x_n} \end{bmatrix}$$

Special Case

As a special case, for $f : \mathbb{R}^n \to \mathbb{R}^1$, the Jacobian is a row vector of dimension $1 \times n$.

Visual Summary
Gradients of Vector-Valued Functions — Jacobian

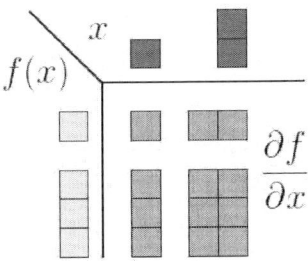

Figure: Summary of dimensions of derivatives.

Example
Gradient of a Vector-Valued Function

We are given
$$\boldsymbol{f}(\boldsymbol{x}) = \boldsymbol{A}\boldsymbol{x}, \quad \boldsymbol{f}(\boldsymbol{x}) \in \mathbb{R}^M, \quad \boldsymbol{A} \in \mathbb{R}^{M \times N}, \quad \boldsymbol{x} \in \mathbb{R}^N.$$

To compute the gradient $\mathrm{d}\boldsymbol{f}/\mathrm{d}\boldsymbol{x}$, we first determine the dimension:
$$\mathrm{d}\boldsymbol{f}/\mathrm{d}\boldsymbol{x} \in \mathbb{R}^{M \times N}.$$

Second, to compute the gradient we determine the partial derivatives of f with respect to every x_j:
$$f_i(\boldsymbol{x}) = \sum_{j=1}^N A_{ij} x_j \implies \frac{\partial f_i}{\partial x_j} = A_{ij}$$

We collect the partial derivatives in the Jacobian and obtain the gradient:
$$\frac{\mathrm{d}\boldsymbol{f}}{\mathrm{d}\boldsymbol{x}} = \begin{bmatrix} \frac{\partial f_1}{\partial x_1} & \cdots & \frac{\partial f_1}{\partial x_N} \\ \vdots & & \vdots \\ \frac{\partial f_M}{\partial x_1} & \cdots & \frac{\partial f_M}{\partial x_N} \end{bmatrix} = \begin{bmatrix} A_{11} & \cdots & A_{1N} \\ \vdots & & \vdots \\ A_{M1} & \cdots & A_{MN} \end{bmatrix} = \boldsymbol{A} \in \mathbb{R}^{M \times N}$$

에듀컨텐츠·휴피아
CH Educontents·Huepia

Chapter 4.
Information Theory

Introduction

In this section, we introduce a few basic concepts from the field of information theory.

Entropy

The entropy of a probability distribution can be interpreted as a measure of uncertainty, or lack of predictability, associated with a random variable drawn from a given distribution.

Entropy in Data Source

We can use entropy to define the information content of a data source. For example, suppose we observe a sequence of symbols $X_n \sim p$ generated from distribution p.

- If p has high entropy, it will be hard to predict the value of each observation X_n.
- By contrast, if p is a degenerate distribution with 0 entropy, then every X_n will be the same, so the dataset does not contain much information.

Entropy for Discrete Random Variables

The entropy of a discrete random variable X with distribution p over K states is defined by

$$\mathbb{H}(X) \triangleq -\sum_{k=1}^{K} p(X = k) \log_2 p(X = k)$$
$$= -\mathbb{E}_X[\log p(X)]$$

Entropy quantifies the expected value of the information contained in a message, usually in units of bits.

Optimal Coding and $\log_2 p(x)$

- Role of $p(x)$: Probability of event x under distribution p.
- Base-2 Logarithm: Relevant for binary codes (0s and 1s).
- Expression $-\log_2 q(x)$:
  - Optimal number of bits for encoding x.
  - Fewer bits for more probable events.

Units and Examples

Usually we use $\log$ base 2, in which case the units are called bits. For example, if $X \in \{1, \ldots, 5\}$ with histogram distribution $p = [0.25, 0.25, 0.2, 0.15, 0.15]$, we find $H = 2.29$ bits. If we use log base e, the units are called nats.

Maximum Entropy

The discrete distribution with maximum entropy is the uniform distribution. For a K-ary random variable, the entropy is maximized if $p(x = k) = 1/K$; in this case,

$$\mathbb{H}(X) = \log_2 K$$

Entropy Visualization

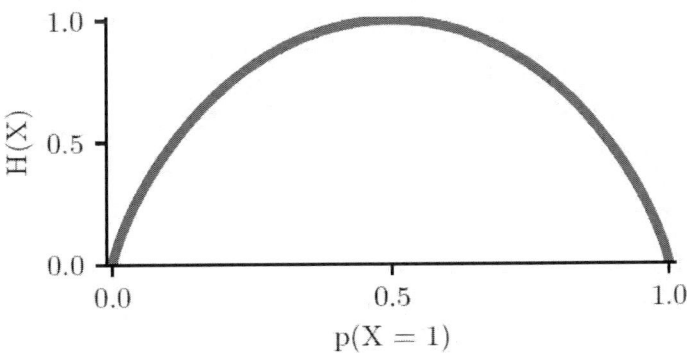

Figure: Entropy of a Bernoulli random variable as a function of θ. The maximum entropy is $\log_2 2 = 1$.

Minimum Entropy

Conversely, the distribution with minimum entropy (which is zero) is any delta-function that puts all its mass on one state. Such a distribution has no uncertainty.

Binary Entropy Function

For binary random variables $X \in \{0, 1\}$, we can write the entropy as:

$$\mathbb{H}(X) = -[\theta \log_2 \theta + (1-\theta)\log_2(1-\theta)]$$

This function is also denoted as $\mathbb{H}(\theta)$.

Fair Coin Example

For a fair coin ($\theta = 0.5$), the entropy reaches its maximum value of 1 bit. A fair coin requires a single yes/no question to determine its state.

Estimating Entropy

Estimating entropy can be challenging when the random variable has many states. For example, if X represents a word in an English document, the long tail of rare words and constant language evolution make it difficult to estimate $\mathbb{H}(X)$ reliably.

Cross Entropy

The cross entropy between two distributions p and q is:

$$\mathbb{H}_{ce}(p, q) = -\sum_{k=1}^{K} p_k \log q_k$$

This measures the expected number of bits needed to encode samples from p using a code based on q.

- "code": a system for representing the possible outcomes in a way that allows for efficient storage or transmission. Less probable events will be mapped to longer strings.
- "code based on distribution q": design the bit sequences according to the probabilities in q, even though the real data is generated according to p. It quantifies how efficient this coding scheme is when the actual data follows distribution p.

Shannon's Source Coding Theorem

The optimal code is obtained when $q = p$, yielding $\mathbb{H}_{ce}(p, p) = \mathbb{H}(p)$. This is known as Shannon's source coding theorem.

Joint Entropy Definition

The joint entropy of random variables X and Y is:

$$\mathbb{H}(X,Y) = -\sum_{x,y} p(x,y) \log_2 p(x,y)$$

Joint Entropy Example

Consider integers n from 1 to 8 with binary variables X and Y defined as follows:

n	1	2	3	4	5	6	7	8
X	0	1	0	1	0	1	0	1
Y	0	1	1	0	1	0	1	0

The joint entropy is $\mathbb{H}(X,Y) = 1.81$ bits.

Bounds on Joint Entropy

If X and Y are independent,
$$\mathbb{H}(X,Y) = \mathbb{H}(X) + \mathbb{H}(Y).$$

Otherwise,
$$\mathbb{H}(X,Y) \leq \mathbb{H}(X) + \mathbb{H}(Y).$$

Lower Bound on Joint Entropy

The lower bound on $\mathbb{H}(X,Y)$ is $\max\{\mathbb{H}(X),\mathbb{H}(Y)\}$ which is always non-negative:
$$\mathbb{H}(X,Y) \geq \max\{\mathbb{H}(X),\mathbb{H}(Y)\} \geq 0$$

$\rightarrow$ Combining variables does not reduce entropy. More unknowns do not reduce uncertainty unless you have additional data.

Conditional Entropy Definition

The conditional entropy of Y given X is defined as the uncertainty in Y after observing X:

$$\mathbb{H}(Y \mid X) \triangleq \mathbb{E}_{p(X)}[\mathbb{H}(p(Y \mid X))]$$
$$= \sum_{x} p(x) \mathbb{H}(p(Y \mid X = x))$$
$$= -\sum_{x,y} p(x,y) \log \frac{p(x,y)}{p(x)}$$
$$= \mathbb{H}(X,Y) - \mathbb{H}(X)$$

Properties of Conditional Entropy

If Y is a deterministic function of X, $\mathbb{H}(Y \mid X) = 0$.
If X and Y are independent, $\mathbb{H}(Y \mid X) = \mathbb{H}(Y)$.

$$\mathbb{H}(Y \mid X) \leq \mathbb{H}(Y)$$

Conditioning on data never increases uncertainty on average.

Definition of Perplexity

The perplexity of a distribution p is:

$$\text{perplexity}(p) \triangleq 2^{\mathbb{H}(p)}$$

Perplexity as a Measure of Predictability

Now suppose we have an empirical distribution based on data $\mathcal{D}$:

$$p_\mathcal{D}(x \mid \mathcal{D}) = \frac{1}{N} \sum_{n=1}^{N} \delta(x - x_n)$$

We can measure how well p predicts $\mathcal{D}$ by computing

$$\text{perplexity}(p_\mathcal{D}, p) \triangleq 2^{\mathbb{H}_{ce}(p_\mathcal{D}, p)}$$

Perplexity in Language Models

In language models, suppose the data is a single long document x of length N, and suppose p is a simple unigram model. In this case, the cross entropy term is given by

$$H = -\frac{1}{N}\sum_{n=1}^{N} \log p(x_n)$$

and hence the perplexity is given by

$$\text{perplexity}(p_\mathcal{D}, p) = 2^H = 2^{-\frac{1}{N}\log\left(\prod_{n=1}^{N} p(x_n)\right)} = \sqrt[N]{\prod_{n=1}^{N} \frac{1}{p(x_n)}}$$

Differential Entropy Definition

For a continuous random variable X with pdf $p(x)$, we define the differential entropy as:

$$h(X) \triangleq -\int_\mathcal{X} p(x) \log p(x) dx$$

Example: Uniform Distribution

Suppose $X \sim U(0, a)$. Then,

$$h(X) = -\int_0^a dx \frac{1}{a} \log \frac{1}{a} = \log a$$

Unlike the discrete case, differential entropy can be negative.

Example: 1D Gaussian Entropy

The entropy of a continuous random variable with probability density function $f(x)$ is defined as:

$$h = -\int_{-\infty}^{\infty} f(x) \log f(x)\, dx$$

For a 1D Gaussian distribution $\mathcal{N}(\mu, \sigma^2)$, the probability density function $f(x)$ is:

$$f(x) = \frac{1}{\sqrt{2\pi\sigma^2}} e^{-\frac{(x-\mu)^2}{2\sigma^2}}$$

After substituting $f(x)$ and simplifying, the integral boils down to:

$$h = \frac{1}{2} \ln\left(2\pi e \sigma^2\right)$$

Introduction

Given two distributions p and q, a divergence measure $D(p, q)$ quantifies how far q is from p.

It is a divergence if $D(p, q) \geq 0$ with equality iff $p = q$.

We focus on the Kullback-Leibler divergence or KL divergence.

Definition for Discrete and Continuous Distributions

For discrete distributions, the KL divergence is:

$$D_{\mathbb{KL}}(p\|q) \triangleq \sum_{k=1}^{K} p_k \log \frac{p_k}{q_k}$$

For continuous distributions, the KL divergence is:

$$D_{\mathbb{KL}}(p\|q) \triangleq \int dx\, p(x) \log \frac{p(x)}{q(x)}$$

Interpreting KL Divergence

We can rewrite the KL divergence as:

$$D_{\mathbb{KL}}(p\|q) = -\mathbb{H}(p) + \mathbb{H}_{ce}(p,q)$$

Here, $-\mathbb{H}(p)$ is the negative entropy, and $\mathbb{H}_{ce}(p,q)$ is the cross entropy. KL divergence can be interpreted as the extra number of bits needed when using distribution q for coding compared to the true distribution p.

KL Divergence for Gaussian Distribution

- Univariate:

$$D_{\mathbb{KL}}\left(\mathcal{N}(x \mid \mu_1, \sigma_1) \| \mathcal{N}(x \mid \mu_2, \sigma_2)\right) = \log \frac{\sigma_2}{\sigma_1} + \frac{\sigma_1^2 + (\mu_1 - \mu_2)^2}{2\sigma_2^2} - \frac{1}{2}$$

- Multivariate:

$$D_{\mathbb{KL}}\left(\mathcal{N}(\boldsymbol{x} \mid \boldsymbol{\mu}_1, \boldsymbol{\Sigma}_1) \| \mathcal{N}(\boldsymbol{x} \mid \boldsymbol{\mu}_2, \boldsymbol{\Sigma}_2)\right)$$
$$= \frac{1}{2}\left[\operatorname{tr}\left(\boldsymbol{\Sigma}_2^{-1}\boldsymbol{\Sigma}_1\right) + (\boldsymbol{\mu}_2 - \boldsymbol{\mu}_1)^\top \boldsymbol{\Sigma}_2^{-1}(\boldsymbol{\mu}_2 - \boldsymbol{\mu}_1) - D + \log\left(\frac{\det(\boldsymbol{\Sigma}_2)}{\det(\boldsymbol{\Sigma}_1)}\right)\right]$$

Jensen's Inequality

To prove that the KL divergence is always non-negative, we use Jensen's inequality, which states that for any convex function f, we have:

$$f\left(\sum_{i=1}^{n} \lambda_i \boldsymbol{x}_i\right) \leq \sum_{i=1}^{n} \lambda_i f(\boldsymbol{x}_i)$$

where $\lambda_i \geq 0$ and $\sum_{i=1}^{n} \lambda_i = 1$.

Information Inequality Theorem

Theorem 17 (Information Inequality)
$D_{\mathbb{KL}}(p\|q) \geq 0$ with equality iff $p = q$.

Proof.

$$\begin{aligned}
-D_{\mathbb{KL}}(p\|q) &= -\sum_{x \in A} p(x) \log \frac{p(x)}{q(x)} \\
&\leq \log \sum_{x \in A} p(x) \frac{q(x)}{p(x)} \\
&\leq \log \sum_{x \in \mathcal{X}} q(x) = \log 1 = 0,
\end{aligned}$$

where $A = \{x : p(x) > 0\}$ is the support of $p(x)$ and $\mathcal{X}$ represents the whole domain space.

Objective in KL Divergence

The goal is to find the distribution q closest to p as measured by KL divergence:

$$q^* = \arg\min_q D_{\mathrm{KL}}(p\|q) = \arg\min_q \int p(x)\log p(x)dx - \int p(x)\log q(x)dx$$

Empirical Distribution

Consider p as the empirical distribution, which focuses on observed training data and has zero mass everywhere else:

$$p_\mathcal{D}(x) = \frac{1}{N}\sum_{n=1}^{N} \delta\left(x - x_n\right)$$

Relation to maximizing likelihood estimation

Using the sifting property of delta functions we get

$$D_{\mathrm{KL}}(p_D \| q) = -\int p_D(x) \log q(x) dx + C$$

$$= -\int \left[\frac{1}{N} \sum_n \delta(x - x_n) \right] \log q(x) dx + C$$

$$= -\frac{1}{N} \sum_n \log q(x_n) + C$$

Minimizing KL divergence to the empirical distribution is equivalent to maximizing likelihood ($\frac{1}{N} \sum_n \log q(x_n)$).

Forward KL or Inclusive KL

- Forward KL:
$$D_{\mathrm{KL}}(p \| q) = \int p(x) \log \frac{p(x)}{q(x)} dx$$

Consider inputs x for which $p(x) > 0$ but $q(x) = 0$. In this case, the term $\log p(x)/q(x)$ will be infinite. Thus minimizing the forward KL will force q to include all the areas of space for which p has non-zero probability.

- Reverse KL:
$$D_{\mathrm{KL}}(q \| p) = \int q(x) \log \frac{q(x)}{p(x)} dx$$

Consider inputs x for which $p(x) = 0$ but $q(x) > 0$. In this case, the term $\log q(x)/p(x)$ will be infinite. Thus minimizing the reverse KL will force q to exclude all the areas of space for which p has zero probability.

Illustration of Mode Covering and Mode Seeking

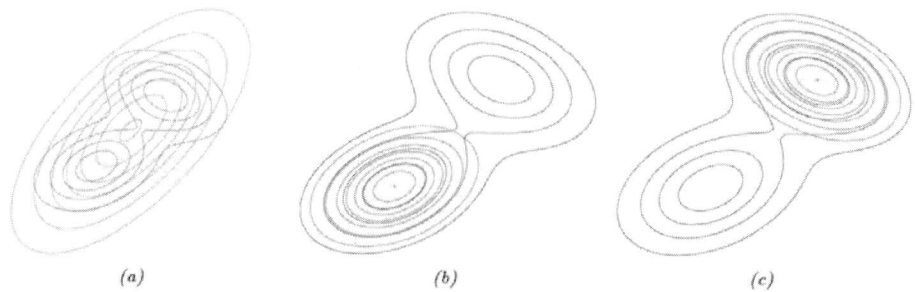

Figure: Illustrating forwards vs reverse KL on a bimodal distribution. The blue curves are the contours of the true distribution p. The red curves are the contours of the unimodal approximation q. (a) Minimizing forwards KL, $D_{\mathbf{KL}}(p\|q)$, wrt q causes q to "cover" p. (b-c) Minimizing reverse KL, $D_{\mathbf{KL}}(q\|p)$ wrt q causes q to "lock onto" one of the two modes of p.

Introduction to Mutual Information

The KL divergence gave us a way to measure how similar two distributions are. How should we measure the dependency between two random variables?

Definition of Mutual Information

The mutual information between random variables X and Y is defined as:

$$\mathbb{I}(X;Y) \triangleq D_{\mathbb{KL}}(p(x,y)\|p(x)p(y)) = \sum_{y \in Y}\sum_{x \in X} p(x,y) \log \frac{p(x,y)}{p(x)p(y)}$$

For continuous random variables, sums are replaced by integrals.

Mutual Information is always non-negative, $\mathbb{I}(X;Y) \geq 0$.

Interpretation of Mutual Information

MI measures the information gain if we update from a model $p(x)p(y)$ to $p(x,y)$.

It can also be re-expressed in terms of entropies:

$$\begin{aligned}\mathbb{I}(X;Y) &= \mathbb{H}(X) - \mathbb{H}(X|Y) \\ &= \mathbb{H}(Y) - \mathbb{H}(Y|X)\end{aligned}$$

→ MI between X and Y: the reduction in uncertainty about X after observing Y, or the reduction in uncertainty about Y after observing X.

Information Diagram

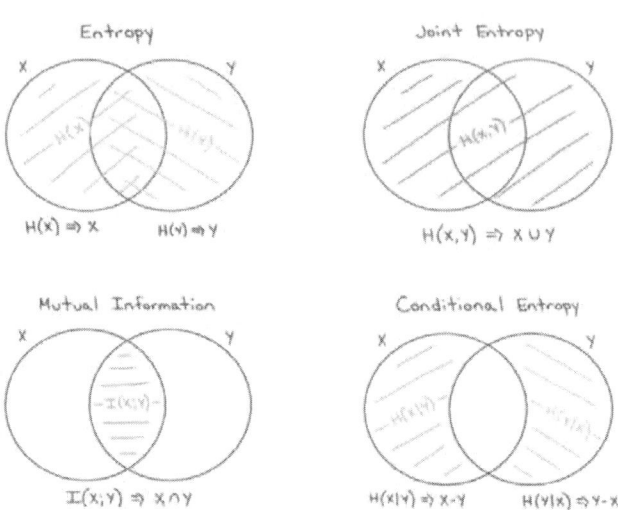

Figure: Information diagram representing mutual information and entropies.

강 의 노 트

1.

2.

3.

Chapter 5.
Statistics

Introduction

In this section, we discuss how to learn parameters from data.

The process of estimating $\boldsymbol{\theta}$ from $\mathcal{D}$ is called *model fitting*, or *training*. The primary objective is:

$$\hat{\boldsymbol{\theta}} = \operatorname*{argmin}_{\boldsymbol{\theta}} \mathcal{L}(\boldsymbol{\theta})$$

Introduction: Continued

The function $\mathcal{L}(\boldsymbol{\theta})$ is generally a loss function. In some cases, closed-form solutions exist, but often optimization algorithms are required.

In addition to a point estimate, $\hat{\boldsymbol{\theta}}$, we also discuss uncertainty quantification, also known as **inference**.

Maximum Likelihood Estimation (MLE)
MLE: Definition

The MLE estimates parameters that maximize the likelihood of observing the training data:

$$\hat{\boldsymbol{\theta}}_{\text{mle}} = \underset{\boldsymbol{\theta}}{\operatorname{argmax}}\ p(\mathcal{D} \mid \boldsymbol{\theta})$$

Under the iid assumption:

$$p(\mathcal{D} \mid \boldsymbol{\theta}) = \prod_{n=1}^{N} p(\boldsymbol{y}_n \mid \boldsymbol{x}_n, \boldsymbol{\theta})$$

Maximum Likelihood Estimation (MLE)
MLE: Log Likelihood

Working with log likelihood simplifies the problem:

$$\ell(\boldsymbol{\theta}) = \sum_{n=1}^{N} \log p(\boldsymbol{y}_n \mid \boldsymbol{x}_n, \boldsymbol{\theta})$$

The MLE is then given by:

$$\hat{\boldsymbol{\theta}}_{\text{mle}} = \underset{\boldsymbol{\theta}}{\operatorname{argmax}} \sum_{n=1}^{N} \log p(\boldsymbol{y}_n \mid \boldsymbol{x}_n, \boldsymbol{\theta})$$

MLE: Negative Log Likelihood (NLL)

Since most optimization algorithms minimize, we can use NLL:

$$\text{NLL}(\boldsymbol{\theta}) = -\sum_{n=1}^{N} \log p\left(\boldsymbol{y}_n \mid \boldsymbol{x}_n, \boldsymbol{\theta}\right)$$

For unsupervised models, the MLE becomes:

$$\hat{\boldsymbol{\theta}}_{\text{mle}} = \operatorname*{argmin}_{\boldsymbol{\theta}} -\sum_{n=1}^{N} \log p\left(\boldsymbol{y}_n \mid \boldsymbol{\theta}\right)$$

Bayesian posterior using a uniform prior

One way for justifying MLE is to view it as a point approximation to the Bayesian posterior $p(\boldsymbol{\theta} \mid \mathcal{D})$ using a uniform prior.

In particular, we approximate the posterior by a delta function:

$$p(\boldsymbol{\theta} \mid \mathcal{D}) = \delta\left(\boldsymbol{\theta} - \hat{\boldsymbol{\theta}}_{\text{map}}\right)$$

where $\hat{\boldsymbol{\theta}}_{\text{map}}$ is the posterior mode, given by:

$$\hat{\boldsymbol{\theta}}_{\text{map}} = \operatorname*{argmax}_{\boldsymbol{\theta}} \log p(\boldsymbol{\theta} \mid \mathcal{D})$$
$$= \operatorname*{argmax}_{\boldsymbol{\theta}} \left(\log p(\mathcal{D} \mid \boldsymbol{\theta}) + \log p(\boldsymbol{\theta})\right)$$

If we use a uniform prior $p(\boldsymbol{\theta}) \propto 1$, then the MAP estimate becomes equal to the MLE:

$$\hat{\boldsymbol{\theta}}_{\text{map}} = \hat{\boldsymbol{\theta}}_{\text{mle}}$$

Predictive Distribution

Another justification for the use of MLE is that the resulting predictive distribution $p\left(y \mid \hat{\theta}_{\text{mle}}\right)$ is as close as possible to the empirical distribution of the data:

$$p_D(y) = \frac{1}{N} \sum_{n=1}^{N} \delta\left(y - y_n\right)$$

Kullback Leibler Divergence

A standard way to measure the (dis)similarity between probability distributions p and q is the Kullback Leibler (KL) divergence. It is defined as:

$$D_{\mathbb{KL}}(p\|q) = \sum_{y} p(y) \log \frac{p(y)}{q(y)}$$
$$= -\mathbb{H}(p) + \mathbb{H}_{ce}(p, q)$$

where $\mathbb{H}(p)$ is the entropy of p, and $\mathbb{H}_{ce}(p, q)$ is the cross-entropy of p and q.

Minimizing KL Divergence

If we define $q(\boldsymbol{y}) = p(\boldsymbol{y} \mid \boldsymbol{\theta})$, and set $p(\boldsymbol{y}) = p_D(\boldsymbol{y})$, then the KL divergence becomes

$$D_{\mathrm{KL}}(p\|q) = -\mathbb{H}(p_D) - \frac{1}{N}\sum_{n=1}^{N} \log p(\boldsymbol{y}_n \mid \boldsymbol{\theta})$$
$$= \mathrm{const} + \mathrm{NLL}(\boldsymbol{\theta})$$

Minimizing the KL is equivalent to minimizing the NLL, which in turn is equivalent to computing the MLE.

Generalization to Supervised Setting

In a supervised setting, we use the following empirical distribution:

$$p_D(\boldsymbol{x}, \boldsymbol{y}) = \frac{1}{N}\sum_{n=1}^{N} \delta(\boldsymbol{x} - \boldsymbol{x}_n)\delta(\boldsymbol{y} - \boldsymbol{y}_n)$$

Bernoulli Distribution Example

> **Example**
>
> Suppose Y is a random variable representing a coin toss, where the event $Y = 1$ corresponds to heads and $Y = 0$ corresponds to tails. Let $\theta = p(Y = 1)$ be the probability of heads. The probability distribution for this random variable is the Bernoulli.

Negative Log-Likelihood for the Bernoulli Distribution

The NLL for the Bernoulli distribution is given by:

$$\begin{aligned}
\text{NLL}(\theta) &= -\log \prod_{n=1}^{N} p(y_n \mid \theta) \\
&= -\log \prod_{n=1}^{N} \theta^{\mathbb{I}(y_n=1)}(1-\theta)^{\mathbb{I}(y_n=0)} \\
&= -\sum_{n=1}^{N} [\mathbb{I}(y_n = 1)\log\theta + \mathbb{I}(y_n = 0)\log(1-\theta)] \\
&= -[N_1 \log\theta + N_0 \log(1-\theta)],
\end{aligned}$$

where

$$N_1 = \sum_{n=1}^{N} \mathbb{I}(y_n = 1), \quad N_0 = \sum_{n=1}^{N} \mathbb{I}(y_n = 0)$$

Computing the MLE

The MLE can be found by solving $\frac{d}{d\theta}\text{NLL}(\theta) = 0$. The derivative of the NLL is:

$$\frac{d}{d\theta}\text{NLL}(\theta) = \frac{-N_1}{\theta} + \frac{N_0}{1-\theta}$$

Thus, the MLE $\hat{\theta}_{\text{mle}}$ is given by:

$$\hat{\theta}_{\text{mle}} = \frac{N_1}{N_0 + N_1}$$

This is simply the empirical fraction of heads, which is an intuitive result.

Introduction to ERM

We can generalize MLE by replacing the (conditional) log loss term with any other loss function:

$$\mathcal{L}(\boldsymbol{\theta}) = \frac{1}{N}\sum_{n=1}^{N} \ell\left(\boldsymbol{y}_n, \boldsymbol{\theta}; \boldsymbol{x}_n\right)$$

This approach is known as Empirical Risk Minimization (ERM).

Issue with MLE and ERM

A problem with MLE and ERM is that they may result in a model that performs well on the training data but poorly on new, unseen data. This phenomenon is known as overfitting.

Example: Predicting Coin Tosses

Consider predicting the probability of heads when tossing a coin $N = 3$ times and observing 3 heads. The MLE would be $\hat{\theta}_{\text{mle}} = 1$, which leads to predicting only heads in the future, an unlikely scenario.

Core Issue and Solution: Regularization

The main solution to overfitting is to use regularization. We add a penalty term to the NLL or empirical risk, resulting in:

$$\mathcal{L}(\boldsymbol{\theta}; \lambda) = \left[\frac{1}{N}\sum_{n=1}^{N} \ell\left(\boldsymbol{y}_n, \boldsymbol{\theta}; \boldsymbol{x}_n\right)\right] + \lambda C(\boldsymbol{\theta})$$

Complexity Penalties

A common penalty is $C(\boldsymbol{\theta}) = -\log p(\boldsymbol{\theta})$. When using log loss, the regularized objective becomes:

$$\mathcal{L}(\boldsymbol{\theta}; \lambda) = -\frac{1}{N}\sum_{n=1}^{N} \log p\left(\boldsymbol{y}_n \mid \boldsymbol{x}_n, \boldsymbol{\theta}\right) - \lambda \log p(\boldsymbol{\theta})$$

MAP Estimation

Maximizing the log posterior is equivalent to minimizing the regularized objective:

$$\hat{\boldsymbol{\theta}} = \underset{\boldsymbol{\theta}}{\operatorname{argmax}} \log p(\boldsymbol{\theta} \mid \mathcal{D}) = \underset{\boldsymbol{\theta}}{\operatorname{argmax}} \left[\log p(\mathcal{D} \mid \boldsymbol{\theta}) + \log p(\boldsymbol{\theta}) - \text{const} \right]$$

This approach is known as MAP (Maximum A Posteriori) estimation.

Example: Coin Tossing Problem

Consider a coin-tossing example where you observe just one head. Using MLE, we get $\theta_{\text{mle}} = 1$, predicting all future coin tosses will result in heads. This can lead to overfitting.

Prior to Avoid Overfitting

To avoid overfitting, a penalty is added to discourage extreme values of θ such as 0 or 1. A beta distribution is used as the prior:

$$p(\theta) = \text{Beta}(\theta \mid a, b) \quad \text{(encouraging } \theta \text{ near to } \frac{a}{a+b} \text{ when } a, b > 1\text{)}$$
$$= \frac{\theta^{a-1}(1-\theta)^{b-1}}{B(a,b)},$$

where $B(a, b)$ is the Beta function, defined as:

$$B(a,b) = \int_0^1 t^{a-1}(1-t)^{b-1}\,dt$$

The log likelihood and log prior are then given by:

$$\ell(\theta) = \log p(\mathcal{D} \mid \theta) + \log p(\theta)$$
$$= [N_1 \log \theta + N_0 \log(1-\theta)] + [(a-1)\log(\theta) + (b-1)\log(1-\theta)]$$

The MAP Estimate

The MAP estimate is given by:

$$\theta_{\text{map}} = \frac{N_1 + a - 1}{N_1 + N_0 + a + b - 2}$$

Introduction to Bayesian Statistics

Bayesian statistics provides a framework for modeling uncertainty using probability distributions rather than point estimates. It is especially important in applications like active learning, avoiding overfitting, and quantifying the trustworthiness of scientific estimates.

Inference in Statistics

Traditional statistical approaches ignore uncertainty in parameter estimates. In Bayesian statistics, uncertainty is modeled using a probability distribution, which is called inference.

Role of Posterior Distribution

Bayesian statistics uses the posterior distribution to represent uncertainty. This approach starts with a prior distribution $p(\theta)$, a likelihood function $p(\mathcal{D} \mid \theta)$, and applies Bayes' rule to compute the posterior distribution:

$$p(\theta \mid \mathcal{D}) = \frac{p(\theta)p(\mathcal{D} \mid \theta)}{p(\mathcal{D})} = \frac{p(\theta)p(\mathcal{D} \mid \theta)}{\int p(\theta')p(\mathcal{D} \mid \theta')d\theta'} \qquad (4)$$

Understanding Marginal Likelihood

The denominator $p(\mathcal{D})$ is called the marginal likelihood. It is obtained by marginalizing over the unknown θ. Although it's a constant independent of θ, it serves as the average probability of the data w.r.t the prior.

However, that $p(\mathcal{D})$ is a constant, independent of θ, so we will often ignore it when we just want to infer the relative probabilities of θ values.

Bayesian Model Averaging (BMA)

After computing the posterior over parameters, the **posterior predictive** distribution over outputs given inputs is calculated by:

$$p(y \mid x, \mathcal{D}) = \int p(y \mid x, \theta) p(\theta \mid \mathcal{D}) d\theta$$

This approach averages over an infinite set of models, reducing the chance of overfitting.

Conjugate Priors

In Bayesian statistics, conjugate priors are a special set of priors that are in the same parameterized family as the posterior. They are important because they allow the posterior to be computed in closed form.

What are Conjugate Priors?

A prior $p(\boldsymbol{\theta}) \in \mathcal{F}$ is a conjugate prior for a likelihood $p(\mathcal{D} \mid \boldsymbol{\theta})$ if the posterior $p(\boldsymbol{\theta} \mid \mathcal{D})$ is also in $\mathcal{F}$.

One example of $\mathcal{F}$ is the exponential family.

The Beta-Binomial Model

The Beta-Binomial model allows us to infer the probability of heads when tossing a coin N times. We use $y_n = 1$ to denote a head and $y_n = 0$ to denote a tail. The data $\mathcal{D}$ contains all the outcomes of N coin tosses.

Bernoulli Likelihood

Assuming the data are independent and identically distributed (iid), the likelihood is given by:

$$p(\mathcal{D} \mid \theta) = \theta^{N_1}(1-\theta)^{N_0} \tag{5}$$

Here, N_1 is the number of heads and N_0 is the number of tails.

The counts N_1 and N_0 are sufficient statistics of the data. This means we only need these counts to infer θ. The total sample size is $N = N_1 + N_0$.

Binomial Likelihood

In the Binomial model, we conduct N trials and observe y heads. The likelihood becomes:

$$p(\mathcal{D} \mid \theta) = \binom{N}{y} \theta^y (1-\theta)^{N-y}$$

Since $\binom{N}{y}$ is independent of θ, it can be ignored, and our inference about θ remains the same as in the Bernoulli model.

Prior Distribution

To simplify the computations, we will assume that the prior $p(\boldsymbol{\theta}) \in \mathcal{F}$ is a conjugate prior for the likelihood function $p(\boldsymbol{y} \mid \boldsymbol{\theta})$. This means that the posterior is in the same parameterized family as the prior, i.e., $p(\boldsymbol{\theta} \mid \mathcal{D}) \in \mathcal{F}$.

To ensure this property when using the Bernoulli (or Binomial) likelihood, we should use a prior of the following form:

$$p(\theta) \propto \theta^{\breve{\alpha}-1}(1-\theta)^{\breve{\beta}-1} \propto \text{Beta}(\theta \mid \breve{\alpha}, \breve{\beta})$$

We recognize this as the pdf of a beta distribution.

Example of Prior Update

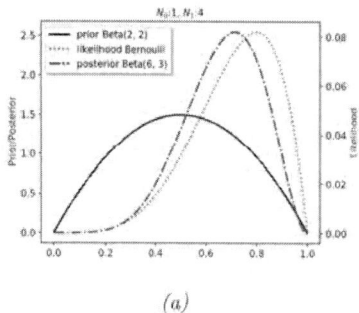 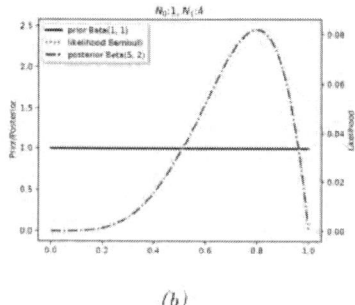

(a) (b)

Figure: Updating a Beta prior with a Bernoulli likelihood with $N_1 = 4, N_0 = 1$.

Posterior Distribution

Multiplying the Bernoulli likelihood with the Beta prior yields a Beta posterior:

$$p(\theta \mid \mathcal{D}) \propto \theta^{N_1}(1-\theta)^{N_0}\theta^{\breve{\alpha}-1}(1-\theta)^{\breve{\beta}-1}$$
$$\propto \text{Beta}\left(\theta \mid \breve{\alpha}+N_1, \breve{\beta}+N_0\right)$$
$$= \text{Beta}(\theta \mid \widehat{\alpha}, \widehat{\beta})$$

where $\widehat{\alpha} \triangleq \breve{\alpha}+N_1$ and $\widehat{\beta} \triangleq \breve{\beta}+N_0$ are the parameters of the posterior. Since the posterior has the same functional form as the prior, we say that the beta distribution is a conjugate prior for the Bernoulli likelihood.

Hyper-Parameters and Pseudo Counts

The prior parameters $\breve{\alpha}$ and $\breve{\beta}$ act as pseudo counts. Posterior parameters can be computed by adding these to the observed counts.

The strength of the prior is determined by $\breve{N} = \breve{\alpha} + \breve{\beta}$, called the equivalent sample size, similar to the observed sample size $N = N_0 + N_1$.

Example: Effect of Different Priors

- For $\breve{\alpha} = \breve{\beta} = 2$, we express a weak preference for $\theta = 0.5$. The effect is shown in Figure 12a.

  The posterior is a "compromise" between prior and likelihood.

- For $\breve{\alpha} = \breve{\beta} = 1$,

$$p(\theta) = \text{Beta}(\theta \mid 1, 1) = \text{Unif}(\theta \mid 0, 1)$$

The effect is shown in Figure 12b.

Posterior shape = Likelihood shape (uninformative prior).

Posterior Mode (MAP Estimate)

The most probable value (MAP estimate) is:

$$\hat{\theta}_{\text{map}} = \arg\max_{\theta} p(\theta \mid \mathcal{D})$$
$$= \arg\max_{\theta} \log p(\theta) + \log p(\mathcal{D} \mid \theta)$$

Using calculus, the MAP is:

$$\hat{\theta}_{\text{map}} = \frac{\breve{\alpha} + N_1 - 1}{\breve{\alpha} + N_1 - 1 + \breve{\beta} + N_0 - 1}$$

Posterior Mean

The posterior mode is a point estimate, which can be less robust. A more robust estimate is the *posterior mean*.

$$\bar{\theta} = \mathbb{E}[\theta \mid \mathcal{D}] = \frac{\widehat{\alpha}}{\widehat{\alpha} + \widehat{\beta}} = \frac{\widehat{\alpha}}{\widehat{N}}$$

where $\widehat{N} = \widehat{\alpha} + \widehat{\beta}$ is the *posterior strength*.

Posterior Mean as a Convex Combination

The posterior mean is a combination of the prior mean and the MLE:

$$\mathbb{E}[\theta \mid \mathcal{D}] = \lambda m + (1 - \lambda)\hat{\theta}_{\mathrm{mle}}$$

where $\lambda = \frac{\breve{N}}{\widehat{N}}$.

This shows the weaker the prior, the closer the posterior mean is to the MLE.

Posterior Variance

To capture uncertainty, the *posterior standard deviation* is used:

$$\text{se}(\theta) = \sqrt{\mathbb{V}[\theta \mid \mathcal{D}]}$$

Variance of the Beta Posterior

The variance of the Beta posterior is:

$$\mathbb{V}[\theta \mid \mathcal{D}] = \frac{\widehat{\alpha}\widehat{\beta}}{(\widehat{\alpha} + \widehat{\beta})^2(\widehat{\alpha} + \widehat{\beta} + 1)}$$

Simplifying for $N \gg \breve{\alpha} + \breve{\beta}$:

$$\mathbb{V}[\theta \mid \mathcal{D}] \approx \frac{\hat{\theta}(1 - \hat{\theta})}{N},$$

where $\hat{\theta}$ is MLE.

Standard Error

The standard error is:

$$\sigma = \sqrt{\mathbb{V}[\theta \mid \mathcal{D}]} \approx \sqrt{\frac{\hat{\theta}(1-\hat{\theta})}{N}}$$

Uncertainty decreases at $1/\sqrt{N}$ and is maximized when $\hat{\theta} = 0.5$.

Posterior Predictive

- Predict future observations using the *plug-in approximation*: First compute an estimate of the parameters based on training data, $\hat{\theta}(\mathcal{D})$, and then to plug that parameter back into the model and use $p(y \mid \theta)$ to predict the future.
- Bernoulli model:

$$p(y = 1 \mid \mathcal{D}) = \int_0^1 p(y = 1 \mid \theta) p(\theta \mid \mathcal{D}) d\theta$$
$$= \int_0^1 \theta \operatorname{Beta}(\theta \mid \widehat{\alpha}, \widehat{\beta}) d\theta = \mathbb{E}[\theta \mid \mathcal{D}] = \frac{\widehat{\alpha}}{\widehat{\alpha} + \widehat{\beta}}$$

Marginal Likelihood

- The marginal likelihood is defined as:

$$p(\mathcal{D} \mid \mathcal{M}) = \int p(\boldsymbol{\theta} \mid \mathcal{M}) p(\mathcal{D} \mid \boldsymbol{\theta}, \mathcal{M}) d\boldsymbol{\theta}$$

- It is important for model selection and hyperparameter estimation.
- Constant wrt $\boldsymbol{\theta}$.

Computation in Beta-Bernoulli Model

Computing marginal likelihood can be hard but in Beta-Bernoulli model, it simplifies:

$$\begin{aligned} p(\mathcal{D}) &= \frac{p(\mathcal{D} \mid \theta) p(\theta)}{p(\theta \mid \mathcal{D})} \\ &= \frac{B(a + N_1, b + N_0)}{B(a, b)} \end{aligned}$$

Beyond Conjugate Priors

We discuss various kinds of priors other than conjugate priors to offer a broader view of Bayesian priors.

Noninformative Priors

- Use when we have little or no domain-specific knowledge.
- Example: Flat prior $p(\mu) \propto 1$ for $\mu \in \mathbb{R}$.

Hierarchical Priors

- Hyperparameters: ξ.
- Multi-level model: $\xi \to \theta \to \mathcal{D}$.

$$p(\xi, \theta, \mathcal{D}) = p(\xi)p(\theta \mid \xi)p(\mathcal{D} \mid \theta)$$

Helps in learning hyperparameters when multiple related parameters need to be estimated.

Empirical Priors

- Computationally convenient approximation.
- Estimate hyperparameters via maximizing the marginal likelihood.

$$\hat{\xi}_{\mathrm{mml}}(\mathcal{D}) = \operatorname*{argmax}_{\xi} \int p(\mathcal{D} \mid \theta)p(\theta \mid \xi)d\theta$$

Hierarchy of Bayesian Methods

Method	Definition
Maximum likelihood	$\hat{\boldsymbol{\theta}} = \mathrm{argmax}_{\boldsymbol{\theta}}\, p(\mathcal{D} \mid \boldsymbol{\theta})$
MAP estimation	$\hat{\boldsymbol{\theta}}(\boldsymbol{\xi}) = \mathrm{argmax}_{\boldsymbol{\theta}}\, p(\mathcal{D} \mid \boldsymbol{\theta}) p(\boldsymbol{\theta} \mid \boldsymbol{\xi})$
ML-II (Empirical Bayes)	$\hat{\boldsymbol{\xi}} = \mathrm{argmax}_{\boldsymbol{\xi}} \int p(\mathcal{D} \mid \boldsymbol{\theta}) p(\boldsymbol{\theta} \mid \boldsymbol{\xi}) d\boldsymbol{\theta}$
MAP-II	$\hat{\boldsymbol{\xi}} = \mathrm{argmax}_{\boldsymbol{\xi}} \int p(\mathcal{D} \mid \boldsymbol{\theta}) p(\boldsymbol{\theta} \mid \boldsymbol{\xi}) p(\boldsymbol{\xi}) d\boldsymbol{\theta}$
Full Bayes	$p(\boldsymbol{\theta}, \boldsymbol{\xi} \mid \mathcal{D}) \propto p(\mathcal{D} \mid \boldsymbol{\theta}) p(\boldsymbol{\theta} \mid \boldsymbol{\xi}) p(\boldsymbol{\xi})$

Note: ML-II is less likely to overfit than regular maximum likelihood.

Bayesian machine learning

So far, we have focused on unconditional models of the form $p(\boldsymbol{y} \mid \boldsymbol{\theta})$. In supervised machine learning, we use conditional models of the form $p(\boldsymbol{y} \mid \boldsymbol{x}, \boldsymbol{\theta})$.

The posterior over the parameters is now $p(\boldsymbol{\theta} \mid \mathcal{D})$, where $\mathcal{D} = \{(\boldsymbol{x}_n, \boldsymbol{y}_n) : n = 1 : N\}$.

This approach is called Bayesian machine learning.

Plugin approximation

Once we have computed the posterior over the parameters, we can compute the posterior predictive distribution as follows:

$$p(y \mid x, \mathcal{D}) = \int p(y \mid x, \theta) p(\theta \mid \mathcal{D}) d\theta$$

Plugin approximation (cont.)

Computing the integral is often intractable. A simple approximation uses the MLE, $\hat{\theta}$ as follows:

$$p(\theta \mid \mathcal{D}) = \delta(\theta - \hat{\theta})$$

where δ is the Dirac delta function. If we use this approximation, then the predictive distribution can be obtained by simply "plugging in" the point estimate into the likelihood:

$$\begin{aligned} p(y \mid x, \mathcal{D}) &= \int p(y \mid x, \theta) p(\theta \mid \mathcal{D}) d\theta \\ &\approx \int p(y \mid x, \theta) \delta(\theta - \hat{\theta}) d\theta \\ &= p(y \mid x, \hat{\theta}) \end{aligned}$$

Approximate Posterior Inference

A variety of methods are available for performing approximate posterior inference.

Example

We will use a beta-Bernoulli model to approximate:

$$p(\theta \mid \mathcal{D}) \propto \left[\prod_{n=1}^{N} \text{Bin}(y_n \mid \theta)\right] \text{Beta}(\theta \mid 1, 1)$$

A Pedagogical Figure

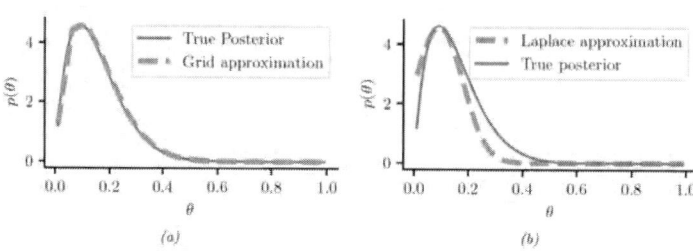

Figure: Approximating the posterior of a beta-Bernoulli model.

Grid approximation

The simplest approach to approximate posterior inference is to partition the space of possible values for the unknowns into a finite set of possibilities, call them $\theta_1, \ldots, \theta_K$, and then to approximate the posterior by brute-force enumeration.

$$p(\theta = \theta_k \mid \mathcal{D}) \approx \frac{p(\mathcal{D} \mid \theta_k) p(\theta_k)}{\sum_{k'=1}^{K} p(\mathcal{D}, \theta_{k'})}$$

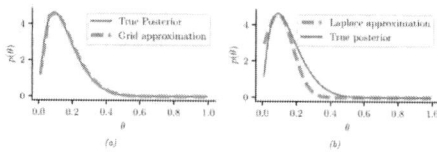

Figure: (a) Grid approximation applied to a 1d problem (b) Laplace approximation applied to a 1d problem

Quadratic (Laplace) approximation

We approximate the posterior using a multivariate Gaussian, known as a Laplace approximation. Suppose we write the posterior as follows:

$$p(\boldsymbol{\theta} \mid \mathcal{D}) = \frac{1}{Z} e^{-\mathcal{E}(\boldsymbol{\theta})}$$

where $\mathcal{E}(\boldsymbol{\theta}) = -\log p(\boldsymbol{\theta}, \mathcal{D})$ is called an energy function, and $Z = p(\mathcal{D})$ is the normalization constant. Performing a Taylor series expansion around the mode $\hat{\boldsymbol{\theta}}$ (i.e., the lowest energy state) we get

$$\mathcal{E}(\boldsymbol{\theta}) \approx \mathcal{E}(\hat{\boldsymbol{\theta}}) + (\boldsymbol{\theta} - \hat{\boldsymbol{\theta}})^\top \boldsymbol{g} + \frac{1}{2}(\boldsymbol{\theta} - \hat{\boldsymbol{\theta}})^\top \mathbf{H}(\boldsymbol{\theta} - \hat{\boldsymbol{\theta}})$$

where g is the gradient at the mode, and $\mathbf{H}$ is the Hessian. Since $\hat{\boldsymbol{\theta}}$ is the mode, the gradient term is zero.

Quadratic (Laplace) approximation (Cont'd)

$$\hat{p}(\boldsymbol{\theta}, \mathcal{D}) = e^{-\mathcal{E}(\hat{\boldsymbol{\theta}})} \exp\left[-\frac{1}{2}(\boldsymbol{\theta} - \hat{\boldsymbol{\theta}})^\mathrm{T} \mathbf{H}(\boldsymbol{\theta} - \hat{\boldsymbol{\theta}})\right]$$

$$\hat{p}(\boldsymbol{\theta} \mid \mathcal{D}) = \frac{1}{Z}\hat{p}(\boldsymbol{\theta}, \mathcal{D}) = \mathcal{N}\left(\boldsymbol{\theta} \mid \hat{\boldsymbol{\theta}}, \mathbf{H}^{-1}\right)$$

$$Z = e^{-\mathcal{E}(\hat{\boldsymbol{\theta}})}(2\pi)^{D/2}|\mathbf{H}|^{-\frac{1}{2}}$$

The last line follows from normalization constant of the multivariate Gaussian.

Variational approximation

Variational inference approximates a complex distribution $p(\boldsymbol{\theta} \mid \mathcal{D})$ with a simpler $q(\boldsymbol{\theta})$ by minimizing a divergence D:

$$q^* = \operatorname*{argmin}_{q \in \mathcal{Q}} D(q, p)$$

Here, $\mathcal{Q}$ is a tractable family (e.g., Gaussian).

Minimizing D as the KL divergence yields the evidence lower bound (ELBO), maximizing which refines the posterior approximation.

Markov Chain Monte Carlo (MCMC) approximation

MCMC is a more flexible, nonparametric approximation method.

$$q(\boldsymbol{\theta}) \approx \frac{1}{S} \sum_{s=1}^{S} \delta\left(\boldsymbol{\theta} - \boldsymbol{\theta}^s\right)$$

The key issue is how to create the posterior samples $\boldsymbol{\theta}^s \sim p(\boldsymbol{\theta} \mid \mathcal{D})$ efficiently, without having to evaluate the normalization constant $p(\mathcal{D}) = \int p(\boldsymbol{\theta}, \mathcal{D}) d\boldsymbol{\theta}$.

Chapter 6.
Inference Overview

Introduction

In the probabilistic approach to machine learning, all unknown quantities are treated as random variables, and endowed with probability distributions. Unknown quantities include

- predictions about the future
- hidden states of a system
- parameters of a model

Introduction (Cont'd)

The process of inference: compute the posterior distribution over unknown quantities (variables), conditioning data. Let $\boldsymbol{\theta}$ represent the unknown variables, and $\mathcal{D}$ represent the known data.

Given a likelihood $p(\mathcal{D} \mid \boldsymbol{\theta})$ and a prior $p(\boldsymbol{\theta})$, we can compute the posterior $p(\boldsymbol{\theta} \mid \mathcal{D})$ using Bayes' rule:

$$p(\boldsymbol{\theta} \mid \mathcal{D}) = \frac{p(\boldsymbol{\theta})p(\mathcal{D} \mid \boldsymbol{\theta})}{p(\mathcal{D})}$$

Introduction (Cont'd)

Once we have the posterior, we can compute posterior expectations of some function of the unknown variables:

$$\mathbb{E}[g(\boldsymbol{\theta}) \mid \mathcal{D}] = \int g(\boldsymbol{\theta}) p(\boldsymbol{\theta} \mid \mathcal{D}) d\boldsymbol{\theta}$$

By defining g appropriately, we can compute many quantities of interest:

$$\begin{aligned}
\text{mean: } & g(\boldsymbol{\theta}) = \boldsymbol{\theta} \\
\text{covariance: } & g(\boldsymbol{\theta}) = (\boldsymbol{\theta} - \mathbb{E}[\boldsymbol{\theta} \mid \mathcal{D}])(\boldsymbol{\theta} - \mathbb{E}[\boldsymbol{\theta} \mid \mathcal{D}])^\mathsf{T} \\
\text{marginals: } & g(\boldsymbol{\theta}) = p(\theta_1 = \theta_1^* \mid \boldsymbol{\theta}_{2:D}) \\
\text{predictive: } & g(\boldsymbol{\theta}) = p(\boldsymbol{y}_{N+1} \mid \boldsymbol{\theta}) \\
\text{expected loss: } & g(\boldsymbol{\theta}) = \ell(\boldsymbol{\theta}, a)
\end{aligned}$$

Marginal Likelihood

If we define $g(\boldsymbol{\theta}) = p(\mathcal{D} \mid \boldsymbol{\theta}, M)$ for model M, then we can also phrase the marginal likelihood as an expectation wrt the prior:

$$\begin{aligned}
\mathbb{E}[g(\boldsymbol{\theta}) \mid M] &= \int g(\boldsymbol{\theta}) p(\boldsymbol{\theta} \mid M) d\boldsymbol{\theta} \\
&= \int p(\mathcal{D} \mid \boldsymbol{\theta}, M) p(\boldsymbol{\theta} \mid M) d\boldsymbol{\theta} \\
&= p(\mathcal{D} \mid M)
\end{aligned}$$

Introduction

Aim of this section

This section gives a high-level summary of algorithmic techniques for computing approximate posteriors and their corresponding expectations. The methods are mostly model-independent.

Common Inference Patterns

Inference Patterns

There are kinds of posteriors we may want to compute, but we identify three main patterns giving rise to different types of inference algorithms.

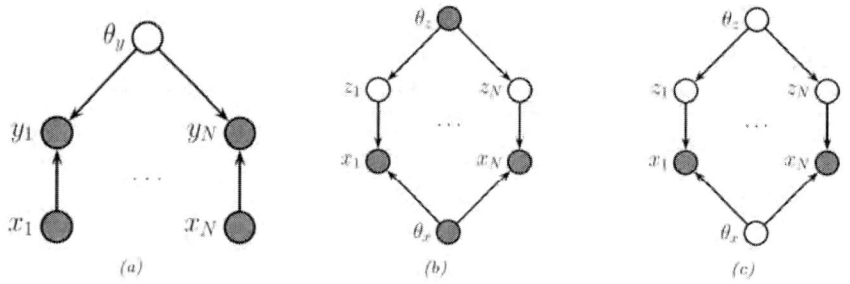

Figure: Graphical models illustrating different types of inference patterns: (a) Global latents (b) Local latents (c) Global and local latents. Shaded nodes are assumed to be known (observed), unshaded nodes are hidden.

Global Latents

- Perform inference in models with global latent variables such as model parameters θ.
- These are shared across all N observed training cases.
- Usual setting for supervised or discriminative learning

The joint distribution has the form:

$$p(y_{1:N}, \theta \mid x_{1:N}) = p(\theta)\left[\prod_{n=1}^{N} p(y_n \mid x_n, \theta)\right]$$

The goal is to compute the posterior $p(\theta \mid x_{1:N}, y_{1:N})$.

Local Latents

- Perform inference in models which have local latent variables, such as hidden states $z_{1:N}$.
- We assume the model parameters θ are known.

The joint distribution has the form

$$p(x_{1:N}, z_{1:N} \mid \theta) = \left[\prod_{n=1}^{N} p(x_n \mid z_n, \theta_x) p(z_n \mid \theta_z)\right]$$

The goal is to compute the posterior $p(z_n \mid x_n, \theta)$ for each n.

Estimating Parameters in Latent Variable Models

- Parameters (θ) often not known in latent variable models (e.g., mixture models).
- One approach: Estimate parameters (e.g., via maximum likelihood) and plug them in.
- Benefit: Latent variables become conditionally independent given θ.
  $\rightarrow$ Allows parallel inference across data.

When θ is unknown: Expectation Maximization (EM)

- In the E-step, infer $p(z_n|x_n, \theta_t)$ for all n.
- In the M-step, update θ_t.
- If exact inference of z_n is not possible, use variational inference (known as variational EM).

When θ is unknown: Minibatch approximation to the likelihood

Mmarginalizing out z_n for each example in the minibatch:

$$\log p(\mathcal{D}_t|\boldsymbol{\theta}_t) = \sum_{n\in\mathcal{D}_t} \log\left[\sum_{z_n} p(x_n, z_n|\boldsymbol{\theta}_t)\right]$$

where $\mathcal{D}_t$ is the minibatch at step t.

- Stochastic Variational Inference (SVI): If marginalization is intractable, use variational inference.
- Amortized SVI: Learn an inference network $q_\phi(z|x;\boldsymbol{\theta})$ where the cost of learning ϕ is amortized across batches.
    - "amortized" here refers to the idea of learning a single, shared inference network $q_\phi(z|x;\boldsymbol{\theta})$ that can be used to perform the inference task for all data points.

Global and Local Latents

- Perform inference in models which have both local and global latent variables.
- Uncertainty in both the local variables z_n and the shared global variables $\boldsymbol{\theta}$

$$p(x_{1:N}, z_{1:N}, \boldsymbol{\theta}) = p(\boldsymbol{\theta}_x)p(\boldsymbol{\theta}_z)\left[\prod_{n=1}^{N} p(x_n \mid z_n, \boldsymbol{\theta}_x) p(z_n \mid \boldsymbol{\theta}_z)\right]$$

Exact Inference Algorithms

- In some cases, we can perform exact posterior inference in a tractable manner.
- This is true if the prior is conjugate to the likelihood. In general, this will be the case when the prior and likelihood are from the same exponential family.

Conjugate Prior

In particular, if the unknown variables are represented by $\boldsymbol{\theta}$, then we assume

$$p(\boldsymbol{\theta}) \propto \exp\left(\boldsymbol{\lambda}_0^\top \mathcal{T}(\boldsymbol{\theta})\right)$$
$$p(\boldsymbol{y}_i \mid \boldsymbol{\theta}) \propto \exp\left(\tilde{\boldsymbol{\lambda}}_i(\boldsymbol{y}_i)^\top \mathcal{T}(\boldsymbol{\theta})\right)$$

where $\mathcal{T}(\boldsymbol{\theta})$ are the sufficient statistics, and $\boldsymbol{\lambda}$ are the natural parameters. We can then compute the posterior by just adding the natural parameters:

$$p(\boldsymbol{\theta} \mid \boldsymbol{y}_{1:N}) = \exp\left(\boldsymbol{\lambda}_*^\top \mathcal{T}(\boldsymbol{\theta})\right)$$
$$\boldsymbol{\lambda}_* = \boldsymbol{\lambda}_0 + \sum_{n=1}^{N} \tilde{\boldsymbol{\lambda}}_n(\boldsymbol{y}_n)$$

Approximate Inference Algorithms

For most probability models, we will not be able to compute marginals or posteriors exactly, so we must resort to using approximate inference. There are many different algorithms, which trade off speed, accuracy, simplicity, and generality.

The MAP Approximation

The simplest approximate inference method is to compute the MAP estimate

$$\hat{\boldsymbol{\theta}} = \operatorname{argmax}_{\boldsymbol{\theta}} p(\boldsymbol{\theta} \mid \mathcal{D}) = \operatorname{argmax}_{\boldsymbol{\theta}} \left[\log p(\boldsymbol{\theta}) + \log p(\mathcal{D} \mid \boldsymbol{\theta})\right]$$

and then to assume that the posterior puts 100% of its probability on this single value:

$$p(\boldsymbol{\theta} \mid \mathcal{D}) \approx \delta(\boldsymbol{\theta} - \hat{\boldsymbol{\theta}})$$

Drawbacks of MAP Approximation

- Provides no measure of uncertainty.
- Plugging in a point estimate can underestimate the predictive uncertainty, which can result in predictions which are not just wrong, but confidently wrong.
- Whereas posterior enables computation of standard errors and credible regions.

Mode: not summarizing distribution

The mode of a posterior distribution is often a very poor choice as a summary statistic, since the mode is usually quite untypical of the distribution, unlike the mean or median.

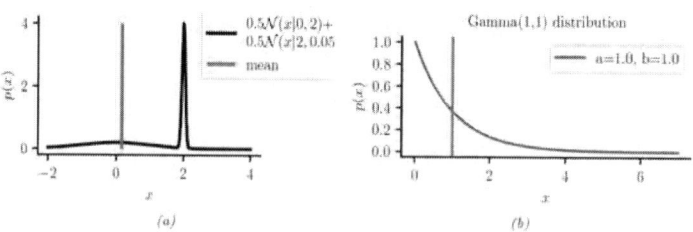

Figure: Two distributions in which the mode (highest point) is untypical of the distribution; the mean (vertical red line) is a better summary. (a) A bimodal distribution. (b) A skewed $Ga(1,1)$ distribution.

MAP and Reparameterization

MAP estimates depend on how we parameterize the probability distribution.

For example, let $\hat{x} = \mathrm{argmax}_x \, p_x(x)$ be the MAP estimate for x and $y = f(x)$ be a transformation of x.

In general it is not the case that $\hat{y} = \mathrm{argmax}_y \, p_y(y)$ is given by $f(\hat{x})$ ($\hat{y} \neq f(\hat{x})$).

Example: MAP and Reparameterization

For $x \sim \mathcal{N}(6, 1)$ and $y = f(x)$ with $f(x) = \frac{1}{1+\exp(-x+5)}$.

By the change of variables:

$$p_y(y) = p_x\left(f^{-1}(y)\right) \left| \frac{df^{-1}(y)}{dy} \right|$$

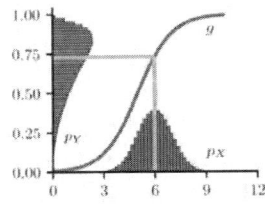

Figure: Example of the transformation of a density under a nonlinear transform. Note that the mode of the transformed distribution is not the transform of the original mode.

Grid Approximation

Partition the space of $\boldsymbol{\theta}$ into regions $r_1, \ldots, r_K$, each representing a region of parameter space of volume Δ centered on $\boldsymbol{\theta}_k$. The probability of being in each region is given by $p\left(\boldsymbol{\theta} \in r_k \mid \mathcal{D}\right) \approx p_k \Delta$, where

$$p_k = \frac{\tilde{p}_k}{\sum_{k'=1}^{K} \tilde{p}_{k'}}, \quad \tilde{p}_k = p\left(\mathcal{D} \mid \boldsymbol{\theta}_k\right) p\left(\boldsymbol{\theta}_k\right)$$

Numerical Approximation

As K increases, grid cell size decreases and the denominator $p(\mathcal{D})$ is just a simple numerical approximation of the integral

$$p(\mathcal{D}) = \int p(\mathcal{D} \mid \boldsymbol{\theta}) p(\boldsymbol{\theta}) d\boldsymbol{\theta} \approx \sum_{k=1}^{K} \Delta \tilde{p}_k$$

Example: Beta-Bernoulli Model

Goal is to approximate $p(\theta \mid \mathcal{D})$.

$$p(\theta \mid \mathcal{D}) \propto \left[\prod_{n=1}^{N} \text{Ber}(y_n \mid \theta)\right] \text{Beta}(1,1)$$

where $\mathcal{D}$ consists of 10 heads and 1 tail (so the total number of observations is $N = 11$), with a uniform prior.

Visualizing Grid Approximation

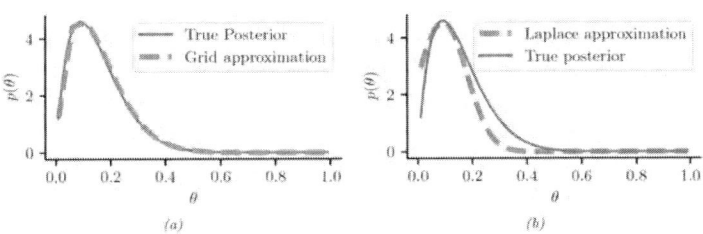

Figure: Approximating the posterior of a beta-Bernoulli model. (a) Grid approximation using 20 grid points. (b) Laplace approximation.

Laplace (quadratic) approximation

Suppose we write the posterior as follows:

$$p(\boldsymbol{\theta} \mid \mathcal{D}) = \frac{1}{Z} e^{-\mathcal{E}(\boldsymbol{\theta})}$$

where $\mathcal{E}(\boldsymbol{\theta}) = -\log p(\boldsymbol{\theta}, \mathcal{D})$ is called an energy function, and $Z = p(\mathcal{D})$ is the normalization constant.

Taylor series expansion around the mode $\hat{\boldsymbol{\theta}}$ (i.e., the lowest energy state):

$$\mathcal{E}(\boldsymbol{\theta}) \approx \mathcal{E}(\hat{\boldsymbol{\theta}}) + (\boldsymbol{\theta} - \hat{\boldsymbol{\theta}})^\top g + \frac{1}{2}(\boldsymbol{\theta} - \hat{\boldsymbol{\theta}})^\top \mathbf{H}(\boldsymbol{\theta} - \hat{\boldsymbol{\theta}})$$

where g is the gradient at the mode, and $\mathbf{H}$ is the Hessian.

Laplace (quadratic) approximation (Cont'd)

Since $\hat{\boldsymbol{\theta}}$ is the mode, the gradient term is zero. Hence

$$\hat{p}(\boldsymbol{\theta}, \mathcal{D}) = e^{-\mathcal{E}(\hat{\boldsymbol{\theta}})} \exp\left[-\frac{1}{2}(\boldsymbol{\theta} - \hat{\boldsymbol{\theta}})^\top \mathbf{H}(\boldsymbol{\theta} - \hat{\boldsymbol{\theta}})\right]$$

$$\hat{p}(\boldsymbol{\theta} \mid \mathcal{D}) = \frac{1}{Z}\hat{p}(\boldsymbol{\theta}, \mathcal{D}) = \mathcal{N}\left(\boldsymbol{\theta} \mid \hat{\boldsymbol{\theta}}, \mathbf{H}^{-1}\right)$$

$$Z = e^{-\mathcal{E}(\hat{\boldsymbol{\theta}})} (2\pi)^{D/2} |\mathbf{H}|^{-\frac{1}{2}}$$

The last line follows from normalization constant of the multivariate Gaussian.

Introduction to Variational Inference (VI)

- Also known as Variational Bayes (VB)
- Optimization-based approach to posterior inference
- More modeling flexibility

Objective of VI

Approximate an intractable probability distribution $(p(\boldsymbol{\theta} \mid \mathcal{D}))$, with one that is tractable, $q(\boldsymbol{\theta})$, so as to minimize some discrepancy D between the distributions:

$$q^* = \operatorname*{argmin}_{q \in \mathcal{Q}} D(q, p)$$

where $\mathcal{Q}$ is some tractable family of distributions (e.g., fully factorized distributions).

Rather than optimizing over functions q, we typically optimize over the parameters of the function q; we denote these variational parameters by ψ.

KL Divergence

It is common to use the KL divergence as the discrepancy measure

$$D(q,p) = D_{\mathbb{KL}}(q(\boldsymbol{\theta} \mid \boldsymbol{\psi}) \| p(\boldsymbol{\theta} \mid \mathcal{D})) = \int q(\boldsymbol{\theta} \mid \boldsymbol{\psi}) \log \frac{q(\boldsymbol{\theta} \mid \boldsymbol{\psi})}{p(\boldsymbol{\theta} \mid \mathcal{D})} d\boldsymbol{\theta}$$

The inference problem then reduces to the following optimization problem:

$$\begin{aligned}
\boldsymbol{\psi}^* &= \operatorname*{argmin}_{\boldsymbol{\psi}} D_{\mathbb{KL}}(q(\boldsymbol{\theta} \mid \boldsymbol{\psi}) \| p(\boldsymbol{\theta} \mid \mathcal{D})) \\
&= \operatorname*{argmin}_{\boldsymbol{\psi}} \mathbb{E}_{q(\boldsymbol{\theta}|\boldsymbol{\psi})} \left[\log q(\boldsymbol{\theta} \mid \boldsymbol{\psi}) - \log \left(\frac{p(\mathcal{D} \mid \boldsymbol{\theta}) p(\boldsymbol{\theta})}{p(\mathcal{D})} \right) \right] \\
&= \operatorname*{argmin}_{\boldsymbol{\psi}} \underbrace{\mathbb{E}_{q(\boldsymbol{\theta}|\boldsymbol{\psi})}[-\log p(\mathcal{D} \mid \boldsymbol{\theta}) - \log p(\boldsymbol{\theta}) + \log q(\boldsymbol{\theta} \mid \boldsymbol{\psi})]}_{-\mathsf{L}(\boldsymbol{\psi})} + \log p(\mathcal{D})
\end{aligned}$$

Evidence Lower Bound (ELBO)

$$\mathsf{L}(\boldsymbol{\psi}) \triangleq \mathbb{E}_{q(\boldsymbol{\theta}|\boldsymbol{\psi})}[\log p(\mathcal{D} \mid \boldsymbol{\theta}) + \log p(\boldsymbol{\theta}) - \log q(\boldsymbol{\theta} \mid \boldsymbol{\psi})]$$

- $\mathsf{L}(\boldsymbol{\psi}) \leq \log p(\mathcal{D}) \because D_{\mathbb{KL}}(q \| p) \geq 0$.
- $\log p(\mathcal{D})$ is called the evidence.
- Maximizing ELBO makes variational posterior closer to the true posterior.

Example: Gaussian

We can choose any kind of approximate posterior. For example, we may use a Gaussian:

$$q(\boldsymbol{\theta} \mid \psi) = \mathcal{N}(\boldsymbol{\theta} \mid \boldsymbol{\mu}, \boldsymbol{\Sigma})$$

- This is different from the Laplace approximation, since in VI, we optimize $\boldsymbol{\Sigma}$, rather than equating it to the Hessian.
- If $\boldsymbol{\Sigma}$ is diagonal, we are assuming the posterior is fully factorized; this is called a mean field approximation.

Automatic differentiation variational inference (ADVI)

- Use change of variables method to convert the parameters to an unconstrained form, and then computes a Gaussian variational approximation.
- Use automatic differentiation to derive the Jacobian term needed to compute the density of the transformed variables.

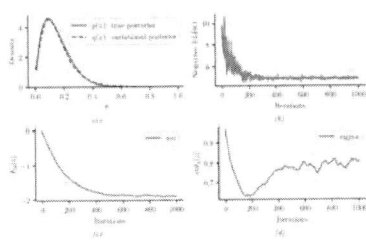

Figure: ADVI applied to the beta-Bernoulli model. (a) Approximate vs true posterior. (b) Negative ELBO over time. (c) Variational μ parameter over time. (d) Variational σ parameter over time.

High Dimensional and Multimodal Posteriors

In many applications, the posterior can be high dimensional and multimodal. Approximating such distributions can be quite challenging.

Comparing Approximation Methods

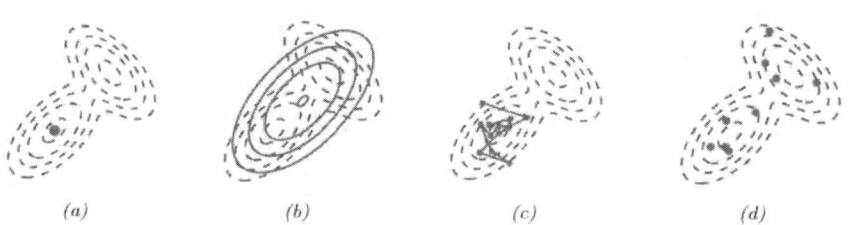

Figure: Different approximations to a bimodal 2d distribution. (a) Local MAP estimate. (b) Parametric Gaussian approximation. (c) (MCMC) Correlated samples from near one mode. (d) Independent samples from the distribution.

Evaluation Criteria

There are many different approximate inference algorithms, each of which make different trade-offs between speed, accuracy, generality, simplicity, etc.

Accuracy vs speed trade-offs can be computed as

$$D_{\mathbb{KL}}\left(p(\boldsymbol{\theta} \mid \mathcal{D}) \| q_t(\boldsymbol{\theta})\right)$$

where $q_t(\boldsymbol{\theta})$ is the approximate posterior after t units of compute time.

Bayesian Risk

It is usually impossible to compute the true posterior $p(\boldsymbol{\theta} \mid \mathcal{D})$. An alternative is to evaluate prediction abilities on out-of-sample data.

More generally, we can compute the Bayesian risk R as follows:

$$R = \mathbb{E}_{p^*(\boldsymbol{x},\boldsymbol{y})}[\ell(\boldsymbol{y}, q(\boldsymbol{y} \mid \boldsymbol{x}, \mathcal{D}))]$$

$$q(\boldsymbol{y} \mid \boldsymbol{x}, \mathcal{D}) = \int p(\boldsymbol{y} \mid \boldsymbol{x}, \boldsymbol{\theta}) q(\boldsymbol{\theta} \mid \mathcal{D}) d\boldsymbol{\theta}$$

에듀컨텐츠·휴피아
CH Educontents·Huepia

Chapter 7.
Variational Inference

Introduction

In this section, we discuss variational inference, which reduces posterior inference to optimization.

Model Components

Consider a model with unknown (latent) variables z, known variables x, and fixed parameters θ.

- Prior: $p_\theta(z)$
- Likelihood: $p_\theta(x \mid z)$
- Unnormalized joint: $p_\theta(x, z) = p_\theta(x \mid z) p_\theta(z)$

Posterior Distribution

The posterior is defined as

$$p_\theta(z \mid x) = \frac{p_\theta(x, z)}{p_\theta(x)}$$

where the normalization constant $p_\theta(x) = \int p_\theta(x, z)dz$ is intractable.

Variational Objective

We seek an approximation $q(z)$ to minimize

$$q = \underset{q \in \mathcal{Q}}{\operatorname{argmin}} D_{\mathbb{KL}}\left(q(z) \| p_\theta(z \mid x)\right)$$

Since we are minimizing over functions (namely distributions q), this is called a variational method.

Variational Objective (Cont'd)

In practice, we choose a parametric family $\mathcal{Q}$ with variational parameters ψ. Objective is to find best variational parameters (for given x) as follows:

$$\begin{aligned}
\psi^* &= \underset{\psi}{\arg\min}\, D_{\mathrm{KL}}\left(q_\psi(z)\|p_\theta(z\mid x)\right) \\
&= \underset{\psi}{\arg\min}\, \mathbb{E}_{q_\psi(z)}\left[\log q_\psi(z) - \log\left(\frac{p_\theta(x\mid z)p_\theta(z)}{p_\theta(x)}\right)\right] \\
&= \underset{\psi}{\arg\min}\, \underbrace{\mathbb{E}_{q_\psi(z)}\left[\log q_\psi(z) - \log p_\theta(x\mid z) - \log p_\theta(z)\right]}_{\mathcal{L}(\theta,\psi\mid x)} + \log p_\theta(x)
\end{aligned}$$

Variational Objective (Cont'd)

The final term $\log p_\theta(x) = \log\left(\int p_\theta(x,z)dz\right)$ is (intractable and) independent of ψ, so we can drop it:

$$\mathcal{L}(\theta,\psi\mid x) = \mathbb{E}_{q_\psi(z)}\left[-\log p_\theta(x,z) + \log q_\psi(z)\right] \qquad (6)$$

Minimizing this objective $\mathcal{L}(\theta,\psi\mid x)$ will minimize the KL divergence, causing our approximation to approach the true posterior.

Illustration of Variational Inference

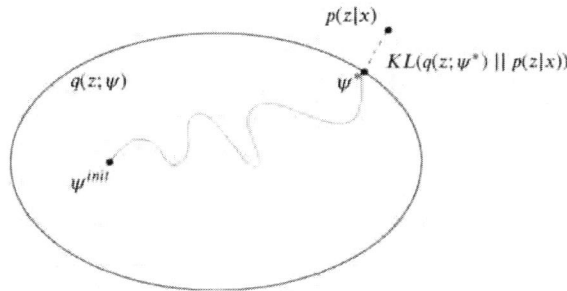

Figure: Illustration of variational inference. The large oval represents the set of variational distributions $\mathcal{Q} = \{q_\psi(z) : \psi \in \Theta\}$, where Θ is the set of possible variational parameters. The true distribution is the point $p(z \mid x)$, which we assume lies outside the set. Our goal is to find the best approximation to p within our variational family; this is the point ψ^* which is closest in KL divergence. We find this point by starting an optimization procedure from the random initial point ψ^{init}.

The View from Physics: Minimize Variational Free Energy

If we define $\mathcal{E}_\theta(z) = -\log p_\theta(z, x)$ as the energy, then the loss in Equation (6) becomes

$$\mathcal{L}(\theta, \psi \mid x) = \mathbb{E}_{q_\psi(z)}\left[\mathcal{E}_\theta(z)\right] - \mathbb{H}(q_\psi) = \text{expected energy - entropy}$$

- Known as the variational free energy (VFE)
- Upper bound on the free energy (FE) : $-\log p_\theta(x)$:

$$D_{\mathrm{KL}}\left(q_\psi(z) \| p_\theta(z \mid x)\right) = \mathcal{L}(\theta, \psi \mid x) + \log p_\theta(x) \geq 0$$
$$\underbrace{\mathcal{L}(\theta, \psi \mid x)}_{\text{VFE}} \geq \underbrace{-\log p_\theta(x)}_{\text{FE}} \quad (7)$$

- Variational inference is equivalent to minimizing the VFE.

The View from Statistics: Maximize the ELBO

The negative of the VFE is known as the evidence lower bound or ELBO function:

$$Ł(\boldsymbol{\theta}, \boldsymbol{\psi} \mid \boldsymbol{x}) \triangleq \mathbb{E}_{q_{\boldsymbol{\psi}}(\boldsymbol{z})} \left[\log p_{\boldsymbol{\theta}}(\boldsymbol{x}, \boldsymbol{z}) - \log q_{\boldsymbol{\psi}}(\boldsymbol{z}) \right] = \text{ELBO}$$

The name "ELBO" arises because

$$Ł(\boldsymbol{\theta}, \boldsymbol{\psi} \mid \boldsymbol{x}) \leq \log p_{\boldsymbol{\theta}}(\boldsymbol{x})$$

where $\log p_{\boldsymbol{\theta}}(\boldsymbol{x})$ is called the "evidence".

Interpretation on ELBO

We can rewrite the ELBO as follows:

$$Ł(\boldsymbol{\theta}, \boldsymbol{\psi} \mid \boldsymbol{x}) = \mathbb{E}_{q_{\boldsymbol{\psi}}(\boldsymbol{z})} \left[\log p_{\boldsymbol{\theta}}(\boldsymbol{x}, \boldsymbol{z}) \right] + \mathbb{H}\left(q_{\boldsymbol{\psi}}(\boldsymbol{z})\right)$$

$$\text{ELBO} = \text{expected log joint} + \text{entropy}$$

or

$$Ł(\boldsymbol{\theta}, \boldsymbol{\psi} \mid \boldsymbol{x}) = \mathbb{E}_{q_{\boldsymbol{\psi}}(\boldsymbol{z})} \left[\log p_{\boldsymbol{\theta}}(\boldsymbol{x} \mid \boldsymbol{z}) + \log p_{\boldsymbol{\theta}}(\boldsymbol{z}) - \log q_{\boldsymbol{\psi}}(\boldsymbol{z}) \right]$$

$$= \mathbb{E}_{q_{\boldsymbol{\psi}}(\boldsymbol{z})} \left[\log p_{\boldsymbol{\theta}}(\boldsymbol{x} \mid \boldsymbol{z}) \right] - D_{\mathbb{KL}} \left(q_{\boldsymbol{\psi}}(\boldsymbol{z}) \| p_{\boldsymbol{\theta}}(\boldsymbol{z}) \right)$$

$$\text{ELBO} = \text{expected log likelihood} - \text{KL from posterior to prior}$$

Choosing the Form of the Variational Posterior

There are two main approaches for choosing the form of the variational posterior, $q_\psi(z \mid x)$.

- Fixed-form VI: pick a convenient functional form, such as multivariate Gaussian, and then optimize the ELBO using gradient-based methods
- Mean field assumption (factored posterior):

$$q_\psi(z) = \prod_{j=1}^{J} q_j(z_j)$$

Parameter Estimation using Variational EM

Estimate model parameters θ by maximizing the log marginal likelihood of the dataset, $\mathcal{D} = \{x_n : n = 1 : N\}$.

$$\log p(\mathcal{D} \mid \theta) = \sum_{n=1}^{N} \log p(x_n \mid \theta)$$

In general, this is intractable to compute.

Latent Variable Model Form

Suppose we have a latent variable model of the form:

$$p(\mathcal{D}, \mathbf{z}_{1:N} \mid \boldsymbol{\theta}) = \prod_{n=1}^{N} p(\mathbf{z}_n \mid \boldsymbol{\theta}) p(\mathbf{x}_n \mid \mathbf{z}_n, \boldsymbol{\theta})$$

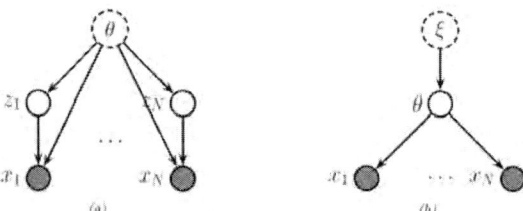

Figure: Graphical models with: (a) Local stochastic latent variables $\mathbf{z}_n$ and global deterministic latent parameter $\boldsymbol{\theta}$. (b) Global stochastic latent parameter $\boldsymbol{\theta}$ and global deterministic latent hyper-parameter $\boldsymbol{\xi}$. The observed variables $\mathbf{x}_n$ are shown by shaded circles.

Local Log Marginal Likelihood

To compute the MLE for $\boldsymbol{\theta}$, we must marginalize out the latent variables to get the local log marginal likelihood:

$$\log p(\mathbf{x}_n \mid \boldsymbol{\theta}) = \log \left[\int p(\mathbf{x}_n \mid \mathbf{z}_n, \boldsymbol{\theta}) p(\mathbf{z}_n \mid \boldsymbol{\theta}) d\mathbf{z}_n \right]$$

Unfortunately, computing this integral is usually intractable.

ELBO as a Lower Bound

The ELBO serves as a lower bound for the log marginal likelihood:

$$Ł(\boldsymbol{\theta}, \boldsymbol{\psi}_n \mid \boldsymbol{x}_n) \leq \log p(\boldsymbol{x}_n \mid \boldsymbol{\theta})$$

We can optimize the model parameters by maximizing:

$$Ł(\boldsymbol{\theta}, \boldsymbol{\psi}_{1:N} \mid \mathcal{D}) \triangleq \sum_{n=1}^{N} Ł(\boldsymbol{\theta}, \boldsymbol{\psi}_n \mid \boldsymbol{x}_n) \leq \log p(\mathcal{D} \mid \boldsymbol{\theta})$$

$\rightarrow$ basis of the variational EM algorithm

Variational EM Algorithm

1. E-step: Maximize the ELBO with respect to the variational parameters $\{\boldsymbol{\psi}_n\}$, yielding $q_{\boldsymbol{\psi}_n}(\boldsymbol{z}_n)$.
2. M-step: Maximize the ELBO (using the new $\boldsymbol{\psi}_n$) with respect to the model parameters $\boldsymbol{\theta}$.

Stochastic Approximation

Sum of the ELBOs for each of the N data samples x_n. Computing this objective can be slow if N is large.

We can replace the objective with a stochastic approximation which is faster to compute:

$$\begin{aligned}&Ł\left(\boldsymbol{\theta}, \boldsymbol{\psi}_{1:N} \mid \mathcal{D}\right) \\&= \sum_{n=1}^{N} Ł\left(\boldsymbol{\theta}, \boldsymbol{\psi}_n \mid \boldsymbol{x}_n\right) \\&\approx \frac{N}{B} \sum_{\boldsymbol{x}_n \in \mathcal{B}} \left[\mathbb{E}_{q_{\psi_n}(\boldsymbol{z}_n)}\left[\log p_{\boldsymbol{\theta}}\left(\boldsymbol{x}_n \mid \boldsymbol{z}_n\right) + \log p_{\boldsymbol{\theta}}\left(\boldsymbol{z}_n\right) - \log q_{\boldsymbol{\psi}_n}\left(\boldsymbol{z}_n\right)\right]\right],\end{aligned}$$

where $B = |\mathcal{B}|$.

SVI Algorithm

This can be used inside a stochastic optimization algorithm such as Stochastic Gradient Descent (SGD). This is called Stochastic Variational Inference (SVI) and allows VI to scale to large datasets.

Inference (or recognition) Network

In SVI, we need to optimize the local variational parameters ψ_n for each example n in the minibatch. This nested optimization can be quite slow.

We can train an inference network to predict ψ_n from the observed data, x_n:

$$\psi_n = f_\phi^{\text{inf}}(x_n)$$

This is known as *amortized* variational inference because the cost of learning ϕ can be amortized across the batches.

ELBO for Amortized VI

The amortized posterior as

$$q(z_n \mid \psi_n) = q\left(z_n \mid f_\phi^{\text{inf}}(x_n)\right) = q_\phi(z_n \mid x_n)$$

The corresponding ELBO becomes

$$\mathcal{L}(\theta, \phi \mid \mathcal{D}) = \sum_{n=1}^{N} \left[\mathbb{E}_{q_\phi(z_n \mid x_n)} \left[\log p_\theta(x_n, z_n) - \log q_\phi(z \mid x_n) \right] \right]$$

Algorithm for Amortized VI

Algorithm Amortized stochastic variational EM

> Initialize θ, ϕ
> **repeat**
>   Sample $x_n \sim p_D$
>   E Step: $\phi = \arg\max_\phi Ł(\theta, \phi | x_n)$
>   M Step: $\theta = \arg\max_\theta Ł(\theta, \phi | x_n)$
> **until** converged

Gradient-based VI

To solve the E-Step and M-Step in variational EM, the gradient descent method can be used.

- E Step:

$$\phi := \phi - \eta \nabla_\phi \mathcal{L}(\phi, \theta | x_n, z_n)$$

- M Step:

$$\theta := \theta - \eta \nabla_\theta \mathcal{L}(\phi, \theta | x_n, z_n)$$

Gradient with Respect to Generative Parameters

The gradient wrt the generative parameters θ is easy because we can push gradients inside the expectation, and use a single Monte Carlo sample:

$$\nabla_{\theta} \mathcal{L}(\theta, \phi \mid x) = \nabla_{\theta} \mathbb{E}_{q_{\phi}(z|x)} [\log p_{\theta}(x, z) - \log q_{\phi}(z \mid x)]$$
$$= \mathbb{E}_{q_{\phi}(z|x)} [\nabla_{\theta} \{\log p_{\theta}(x, z) - \log q_{\phi}(z \mid x)\}]$$
$$\approx \nabla_{\theta} \log p_{\theta}(x, z^s)$$

where $z^s \sim q_{\phi}(z \mid x)$. This is an unbiased estimate of the gradient and can be used with SGD.

Gradient with Respect to Inference Parameters

The gradient wrt the inference parameters ϕ is harder to compute:

$$\nabla_{\phi} \mathcal{L}(\theta, \phi \mid x) \neq \mathbb{E}_{q_{\phi}(z|x)} [\nabla_{\phi} \{\log p_{\theta}(x, z) - \log q_{\phi}(z \mid x)\}]$$

However, we can often use the reparameterization trick or blackbox VI.

Reparameterized VI

- Reparameterization trick: enables to take gradients wrt distributions over continuous latent variables $z \sim q_\phi(z \mid x)$
- Key idea: Rewrite the random variable $z \sim q_\phi(z \mid x)$ as some differentiable (and invertible) transformation g of another random variable $\epsilon \sim p(\epsilon)$, which does not depend on ϕ, i.e., we assume we can write

$$z = g(\phi, x, \epsilon)$$

- Example - Gaussian distribution:

$$z \sim \mathcal{N}(\mu, \mathrm{diag}(\sigma)) \iff z = \mu + \epsilon \odot \sigma, \epsilon \sim \mathcal{N}(0, I)$$

Reparameterization Trick Example

Using this, we have

$$\mathbb{E}_{q_\phi(z|x)}[f(z)] = \mathbb{E}_{p(\epsilon)}[f(z)] \text{ s.t. } z = g(\phi, x, \epsilon)$$

where we define

$$f_{\theta,\phi}(z) = \log p_\theta(x, z) - \log q_\phi(z \mid x)$$

Hence

$$\nabla_\phi \mathbb{E}_{q_\phi(z|x)}[f(z)] = \nabla_\phi \mathbb{E}_{p(\epsilon)}[f(z)] = \mathbb{E}_{p(\epsilon)}[\nabla_\phi f(z)]$$

→ We can approximate with a single Monte Carlo sample and this lets us prpagate gradients back throught the f function

Reparameterization Trick Illustration

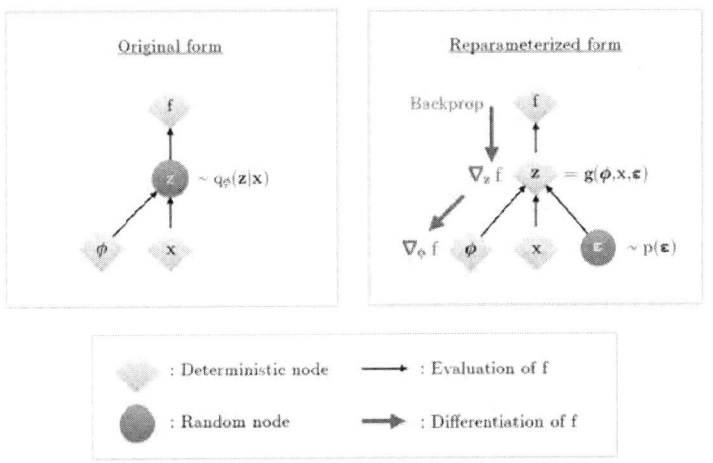

Figure: Illustration of the reparameterization trick

Change of Variables Formula

To compute $\log q_\phi(z \mid x)$, we need:

$$\log q_\phi(z \mid x) = \log p(\epsilon) - \log \left| \det \left(\frac{\partial z}{\partial \epsilon} \right) \right| \tag{8}$$

where the Jacobian matrix $\frac{\partial z}{\partial \epsilon}$ is:

$$\frac{\partial z}{\partial \epsilon} = \begin{pmatrix} \frac{\partial z_1}{\partial \epsilon_1} & \cdots & \frac{\partial z_1}{\partial \epsilon_k} \\ \vdots & \ddots & \vdots \\ \frac{\partial z_k}{\partial \epsilon_1} & \cdots & \frac{\partial z_k}{\partial \epsilon_k} \end{pmatrix}$$

We design the transformation g such that this Jacobian is easy to compute.

Example: Gaussian with Diagonal Covariance (Mean Field)

Suppose we use a fully factorized Gaussian posterior. Then the reparameterization process becomes:

$$\epsilon \sim \mathcal{N}(0, \mathbf{I})$$
$$\mathbf{z} = \boldsymbol{\mu} + \boldsymbol{\sigma} \odot \epsilon$$

where the inference network generates the parameters of the transformation:

$$(\boldsymbol{\mu}, \log \boldsymbol{\sigma}) = f_\phi^{\text{inf}}(\mathbf{x})$$

Evaluating the ELBO

To sample from the posterior $q_\phi(\mathbf{z} \mid \mathbf{x})$, we sample $\epsilon \sim \mathcal{N}(0, \mathbf{I})$, and then compute $\mathbf{z}$ using the parameters generated by the inference network.

Given the sample, we need to evaluate the ELBO:

$$f(\mathbf{z}) = \log p_{\boldsymbol{\theta}}(\mathbf{x} \mid \mathbf{z}) + \log p_{\boldsymbol{\theta}}(\mathbf{z}) - \log q_\phi(\mathbf{z} \mid \mathbf{x})$$

Evaluating $\log p_\theta(x \mid z)$, q_ϕ

- $\log p_\theta(x \mid z)$: just plug z into the likelihood
- $\log q_\phi(z \mid x)$: need to use the change of variables formula from Equation (8). The Jacobian is given by $\frac{\partial z}{\partial \epsilon} = \text{diag}(\sigma)$. Hence

$$\log q_\phi(z \mid x) = \sum_{k=1}^{K} [\log \mathcal{N}(\epsilon_k \mid 0, 1) - \log \sigma_k]$$

$$= -\sum_{k=1}^{K} \left[\frac{1}{2} \log(2\pi) + \frac{1}{2} \epsilon_k^2 + \log \sigma_k \right]$$

Evaluating $p(z)$

Use the transformation $z = 0 + 1 \odot \epsilon$.

The Jacobian is the identity and we get

$$\log p(z) = \sum_{k=1}^{K} \left[\frac{1}{2} z_k^2 + \frac{1}{2} \log(2\pi) \right] \qquad (9)$$

Reparameterization for Full Covariance

Now consider using a full covariance Gaussian posterior. We compute a Cholesky decomposition of the covariance:

$$\Sigma = LL^\top$$

where L is a lower triangular matrix with non-zero entries on the diagonal. The reparameterization becomes:

$$\epsilon \sim \mathcal{N}(0, I)$$
$$z = \mu + L\epsilon$$

Jacobian of Transformation

The Jacobian of the affine transformation is given by:

$$\frac{\partial z}{\partial \epsilon} = L$$

Its determinant is:

$$\log \left| \det \frac{\partial z}{\partial \epsilon} \right| = \sum_{k=1}^{K} \log |L_{kk}|$$

Construction of L

We can compute $\mathbf{L}$ using the formula:

$$\mathbf{L} = \mathbf{M} \odot \mathbf{L}' + \text{diag}(\boldsymbol{\sigma})$$

where $\mathbf{M}$ is a masking matrix with 0s on and above the diagonal, and 1s below the diagonal. The parameters $(\boldsymbol{\mu}, \log \boldsymbol{\sigma}, \mathbf{L}')$ are predicted by the inference network.

Determinant of Jacobian with σ

With this construction, the diagonal entries of $\mathbf{L}$ are given by $\boldsymbol{\sigma}$. Hence:

$$\log \left| \det \frac{\partial \mathbf{z}}{\partial \boldsymbol{\epsilon}} \right| = \sum_{k=1}^{K} \log |L_{kk}| = \sum_{k=1}^{K} \log \sigma_k$$

MLE for Latent Variable Models (LVMs)

- Focus on reparameterized SVI.
- Specifically used for continuous latent variables.
- Amortized inference: learning an inference network with parameters ϕ.
- Prediction of local variational parameters ψ_n given x_n.

Reparameterized Amortized SVI for MLE of LVM

Algorithm Reparameterized Amortized SVI for MLE of an LVM

Initialize θ, ϕ
repeat
    Sample $x_n \sim p_D$
    Sample $\epsilon \sim q_0$
    Compute $z_n = g(\phi, x_n, \epsilon_n)$
    Compute $\mathcal{L}(\theta, \phi | x_n, z_n) = -\log p_\theta(x_n, z_n) + \log q_\phi(z_n | x_n)$
    Update $\theta := \theta - \eta \nabla_\theta \mathcal{L}(\phi, \theta | x_n, z_n)$
    Update $\phi := \phi - \eta \nabla_\phi \mathcal{L}(\phi, \theta | x_n, z_n)$
until converged

Blackbox Variational Inference (BBVI)

- $\tilde{\mathcal{L}}(\boldsymbol{\psi}, \boldsymbol{z}) = \log p(\boldsymbol{z}, \boldsymbol{x}) - \log q_{\boldsymbol{\psi}}(\boldsymbol{z})$
- Assume we can evaluate $\tilde{\mathcal{L}}(\boldsymbol{\psi}, \boldsymbol{z})$ pointwise.
- Do not assume availability of gradients.
  $\rightarrow$ Estimate the gradient of the ELBO

Estimating the Gradient using REINFORCE

Rewrite the ELBO as

$$\mathsf{L}(\boldsymbol{\psi}) = \mathbb{E}_{q(\boldsymbol{z}|\boldsymbol{\psi})}[\tilde{\mathcal{L}}(\boldsymbol{\psi}, \boldsymbol{z})]$$
$$= \mathbb{E}_{q(\boldsymbol{z}|\boldsymbol{\psi})}[\log p(\boldsymbol{x}, \boldsymbol{z}) - \log q(\boldsymbol{z} \mid \boldsymbol{\psi})]$$

Gradient of ELBO I

The ELBO $\mathcal{L}(\psi)$ is given by an expectation:

$$\mathcal{L}(\psi) = \mathbb{E}_{q(z|\psi)}[\log p(x,z) - \log q(z \mid \psi)]$$

To differentiate this with respect to ψ, we get:

$$\nabla_\psi \mathcal{L}(\psi) = \nabla_\psi \int q(z \mid \psi)(\log p(x,z) - \log q(z \mid \psi))\, dz$$

Applying the chain rule, the derivative of the integral (which is essentially an expectation) becomes:

$$\nabla_\psi \mathcal{L}(\psi) = \int \nabla_\psi q(z \mid \psi)(\log p(x,z) - \log q(z \mid \psi))\, dz$$

Gradient of ELBO II

To simplify the expression, we rewrite $\nabla_\psi q(z \mid \psi)$ as $q(z \mid \psi)\nabla_\psi \log q(z \mid \psi)$:

$$\nabla_\psi \mathcal{L}(\psi) = \int q(z \mid \psi)\nabla_\psi \log q(z \mid \psi)(\log p(x,z) - \log q(z \mid \psi))\, dz$$

Finally, recognizing that this is an expectation, we rewrite it as:

$$\nabla_\psi \mathcal{L}(\psi) = \mathbb{E}_{q(z|\psi)}[\tilde{\mathcal{L}}(\psi,z)\nabla_\psi \log q(z \mid \psi)]$$

→ We only require that the sampling distribution is differentiable, not the objective itself.

Monte Carlo Approximation

We can then compute a Monte Carlo approximation to this:

$$\widehat{\nabla_\psi \mathcal{L}(\psi_t)} = \frac{1}{S} \sum_{s=1}^{S} \tilde{\mathcal{L}}(\psi, z_s) \nabla_\psi \log q_\psi(z_s) \Big|_{\psi=\psi_t} \quad (10)$$

- Can be optimized using gradient-based methods like SGD or Adam.

강의노트

1.

2.

3.

Chapter 8.
Variational Autoencoder

Introduction

In this section, we discuss generative models of the form

$$z \sim p_\theta(z)$$
$$x \mid z \sim \text{Expfam}\left(x \mid d_\theta(z)\right)$$

Introduction (Cont'd)

The model is called a deep latent variable model (DLVM) or deep latent Gaussian model (DLGM) depending on the prior.

- Posterior inference is computationally intractable, as is computing the marginal likelihood

$$p_\theta(x) = \int p_\theta(x \mid z) p_\theta(z) dz$$

- Need for approximate inference methods.

Amortized Inference

Use a recognition network (or inference network) that predicts ψ_n from the observed data, x_n:

$$\psi_n = f_\phi^{\text{inf}}(x_n)$$

where ψ is variational parameters.

The amortized posterior as

$$q(z_n \mid \psi_n) = q\left(z_n \mid f_\phi^{\text{inf}}(x_n)\right) = q_\phi(z_n \mid x_n)$$

Schematic Illustration of a VAE

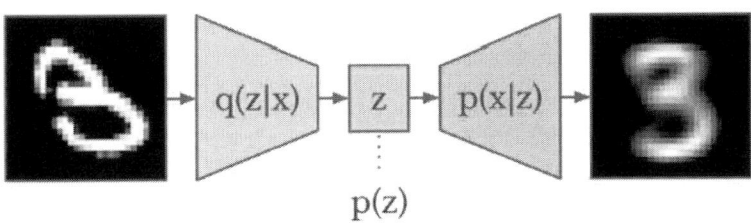

Figure: Schematic illustration of a VAE.

Modeling Assumptions

In the simplest setting, a VAE defines a generative model:

$$p_{\boldsymbol{\theta}}(\boldsymbol{z}, \boldsymbol{x}) = p_{\boldsymbol{\theta}}(\boldsymbol{z}) p_{\boldsymbol{\theta}}(\boldsymbol{x} \mid \boldsymbol{z})$$

where $p_{\boldsymbol{\theta}}(\boldsymbol{z})$ is usually a Gaussian, and $p_{\boldsymbol{\theta}}(\boldsymbol{x} \mid \boldsymbol{z})$ is usually a product of exponential family distributions (e.g., Gaussians or Bernoullis), with parameters computed by a neural network decoder, $d_{\boldsymbol{\theta}}(\boldsymbol{z})$.

Generative Model (Continued)

For example, for binary observations, we can use:

$$p_{\boldsymbol{\theta}}(\boldsymbol{x} \mid \boldsymbol{z}) = \prod_{d=1}^{D} \mathrm{Ber}\left(x_d \mid \sigma\left(d_{\boldsymbol{\theta}}(\boldsymbol{z})\right)\right)$$

Recognition Model

A VAE also fits a recognition model to perform approximate posterior inference:

$$q_\phi(z \mid x) = q\left(z \mid e_\phi(x)\right) \approx p_\theta(z \mid x)$$

Here $q_\phi(z \mid x)$ is usually a Gaussian, with parameters computed by a neural network encoder $e_\phi(x)$:

$$q_\phi(z \mid x) = \mathcal{N}(z \mid \mu, \mathrm{diag}(\exp(\ell)))$$
$$(\mu, \ell) = e_\phi(x)$$

where $\ell = \log \sigma$.

Amortized Inference

The idea of training an inference network to "invert" a generative network is called amortized inference.

- First proposed in the Helmholtz machine.
- The VAE optimizes a variational lower bound on the log-likelihood.

Model Fitting for VAEs

Fitting a VAE using amortized stochastic variational inference involves

- Likelihood models for $p_\theta(x \mid z)$
- Posterior approximations for $q_\phi(z \mid x)$

Reminder: Reparameterized Amortized SVI

Algorithm Reparameterized Amortized SVI for MLE of an LVM

    Initialize θ, ϕ
    repeat
        Sample $x_n \sim p_D$
        Sample $\epsilon \sim q_0$
        Compute $z_n = g(\phi, x_n, \epsilon_n)$
        Compute $\mathcal{L}(\theta, \phi | x_n, z_n) = -\log p_\theta(x_n, z_n) + \log q_\phi(z_n | x_n)$
        Update $\theta := \theta - \eta \nabla_\theta \mathcal{L}(\phi, \theta | x_n, z_n)$
        Update $\phi := \phi - \eta \nabla_\phi \mathcal{L}(\phi, \theta | x_n, z_n)$
    until converged

Fitting Algorithm

Algorithm Fitting a VAE with Bernoulli likelihood and full covariance Gaussian posterior.

1: Initialize θ, ϕ
2: **repeat**
3:     Sample $x_n \sim p_D$
4:     Sample $\epsilon \sim q_0$
5:     $(\mu, \log \sigma, L') = e_{\phi(x)}$
6:     $M = \mathrm{np.triu}(\mathrm{np.ones}(K), -1)$
7:     $L = M \odot L' + \mathrm{diag}(\sigma)$
8:     $z = L\epsilon + \mu$
9:     $p = d_\theta(z)$
10:    Compute $\mathcal{L}_{\log qz}, \mathcal{L}_{\log pz}, \mathcal{L}_{\log px}$
11:    Update θ and ϕ
12: **until** converged

$\mathcal{L}_{\log qz}$

Things to know for deriving $\mathcal{L}_{\log qz}$:

- Change of variables: $\log q_\phi(z \mid x) = \log p(\epsilon) - \log \left|\det \left(\frac{\partial z}{\partial \epsilon}\right)\right|$
- Reparameterization:

$$z \sim \mathcal{N}(\mu, \Sigma),\ \Sigma = LL^\top$$
$$\epsilon \sim \mathcal{N}(0, I)$$
$$z = \mu + L\epsilon$$
$$\frac{\partial z}{\partial \epsilon} = L,\ \log\left|\det \frac{\partial z}{\partial \epsilon}\right| = \sum_{k=1}^{K} \log |L_{kk}|$$

Because L can be computed $L = M \odot L' + \mathrm{diag}(\sigma)$,
$\log\left|\det \frac{\partial z}{\partial \epsilon}\right| = \sum_{k=1}^{K} \log |L_{kk}| = \sum_{k=1}^{K} \log \sigma_k$, where M is a masking matrix with 0s on and above the diagonal, and 1s below the diagonal.

$\mathcal{L}_{\log qz}$ (Cont'd)

Therefore

- $\log p(\boldsymbol{\epsilon}) = \sum_{k=1}^{K} [\log \mathcal{N}(\epsilon_k \mid 0, 1)]$
- $-\log \left|\det \left(\frac{\partial \boldsymbol{z}}{\partial \boldsymbol{\epsilon}}\right)\right| = -\sum_{k=1}^{K} \log \sigma_k$

Using these, we arrive at

$$\mathcal{L}_{\log qz} = -\sum_{k=1}^{K} \left[\frac{1}{2}\epsilon_k^2 + \frac{1}{2}\log(2\pi) + \log \sigma_k\right]$$

$\mathcal{L}_{\log pz}$

Assuming $p(\boldsymbol{z}) = \mathcal{N}(\boldsymbol{z} \mid 0, \boldsymbol{I})$, $\boldsymbol{z} = 0 + 1 \odot \boldsymbol{\epsilon}$, so the Jacobian is the identity and we get

$$\mathcal{L}_{\log pz} = -\log p(\boldsymbol{z}) = -\sum_{k=1}^{K} \left[\frac{1}{2}z_k^2 + \frac{1}{2}\log(2\pi)\right]$$

$\mathcal{L}_{\log px}$

Based on Bernoulli likelihood

$$\mathcal{L}_{\log px} = -\log p_{\boldsymbol{\theta}}(\boldsymbol{x} \mid \boldsymbol{z}) = -\sum_{d=1}^{D} [x_d \log p_d + (1 - x_d) \log(1 - p_d)]$$

where $\boldsymbol{p} = d_\theta(\boldsymbol{z})$.

If Gaussian likelihood is used

$$\mathcal{L}_{\log px} = -\log p_{\boldsymbol{\theta}}(\boldsymbol{x} \mid \boldsymbol{z}) = -\frac{1}{2\sigma^2} \|\boldsymbol{x} - \boldsymbol{p}\|_2^2 + \text{const}$$

VAEs vs Autoencoders

Variational Autoencoders (VAEs) are closely related to deterministic Autoencoders (AEs). The primary differences are:

- AEs aim to maximize the log likelihood of the reconstruction without any KL term.
- Encoding in AEs is deterministic.

The encoding in VAEs computes not just $\mathbb{E}[\boldsymbol{z} \mid \boldsymbol{x}]$, but also $\mathbb{V}[\boldsymbol{z} \mid \boldsymbol{x}]$.

VAEs vs Autoencoders (Cont'd)

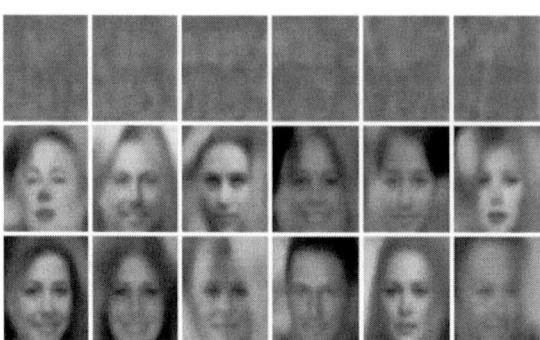

Figure: When fed with random $z \sim \mathcal{N}(0, I)$. Row 1: AE, Row 2: β-VAE ($\beta = 0.5$), Row 3: VAE ($\beta = 1$)

VAEs are advantageous over AEs in that they define a proper generative model. Autoencoder only knows how to decode latent codes derived from the training set, so does poorly when fed random inputs.

VAEs vs Autoencoders (Cont'd)

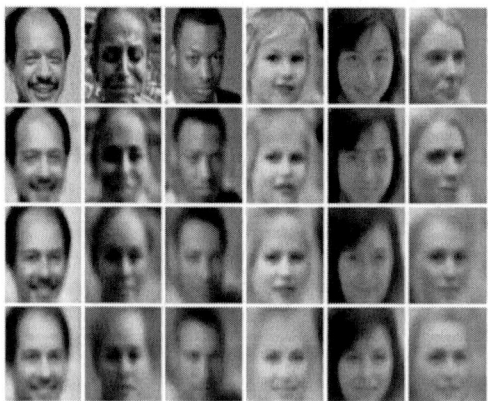

Figure: Image reconstruction comparison. Row 1: original images, Row 2: AE, Row 3: β-VAE ($\beta = 0.5$), Row 4: VAE ($\beta = 1$)

Both models can reconstruct input images but VAEs tend to be somewhat blurry.

Advantages and Drawbacks

- VAEs can generate new, sensible-looking images.
- AEs can only decode latent codes from the training set.
- VAEs tend to generate blurrier reconstructions.

Notations for Alternative Expressions for the ELBO

- Joint generative distribution:
$$p_\theta(x, z) = p_\theta(z)p_\theta(x \mid z)$$

- Generative data marginal:
$$p_\theta(x) = \int_z p_\theta(x, z)dz$$

- Generative posterior:
$$p_\theta(z \mid x) = \frac{p_\theta(x, z)}{p_\theta(x)}$$

- Joint inference distribution:
$$q_{\mathcal{D},\phi}(z, x) = p_\mathcal{D}(x)q_\phi(z \mid x)$$

where $p_\mathcal{D}(x) = \frac{1}{N}\sum_{n=1}^{N}\delta(x_n - x)$ is the empirical distribution.

Notations for Alternative Expressions for the ELBO

- Inference latent marginal (or aggregated posterior):

$$q_{\mathcal{D},\phi}(z) = \int_x q_{\mathcal{D},\phi}(x,z)dx$$

- Inference likelihood:

$$q_{\mathcal{D},\phi}(x \mid z) = \frac{q_{\mathcal{D},\phi}(x,z)}{q_{\mathcal{D},\phi}(z)}$$

Alternative Versions of the ELBO

The ELBO averaged over all the data is given by

$$\begin{aligned}
\mathcal{L}(\theta,\phi \mid \mathcal{D}) &= \mathbb{E}_{p_{\mathcal{D}}(x)}\left[\mathbb{E}_{q_\phi(z\mid x)}[\log p_\theta(x \mid z)]\right] - \mathbb{E}_{p_{\mathcal{D}}(x)}[D_{\mathrm{KL}}(q_\phi(z \mid x) \| p_\theta(z))] \\
&= \mathbb{E}_{q_{\mathcal{D},\phi}(x,z)}[\log p_\theta(x \mid z) + \log p_\theta(z) - \log q_\phi(z \mid x)] \\
&= \mathbb{E}_{q_{\mathcal{D},\phi}(x,z)}\left[\log \frac{p_\theta(x,z)}{q_{\mathcal{D},\phi}(x,z)} + \log p_{\mathcal{D}}(x)\right] \\
&= -D_{\mathrm{KL}}(q_{\mathcal{D},\phi}(x,z) \| p_\theta(x,z)) + \mathbb{E}_{p_{\mathcal{D}}(x)}[\log p_{\mathcal{D}}(x)] \\
&\stackrel{c}{=} -D_{\mathrm{KL}}(q_\phi(x,z) \| p_\theta(x,z)) \\
&\stackrel{c}{=} -D_{\mathrm{KL}}(p_{\mathcal{D}}(x) \| p_\theta(x)) - \mathbb{E}_{p_{\mathcal{D}}(x)}[D_{\mathrm{KL}}(q_\phi(z \mid x) \| p_\theta(z \mid x))],
\end{aligned}$$

where $\stackrel{c}{=}$ to mean equal up to additive constants.

→ Second KL term is minimized by fitting the true posterior. Thus if the posterior family is limited, there may be a conflict between these objectives.

Alternative Versions of the ELBO (Cont'd)

The ELBO can also be written as

$$\mathrm{E}(\boldsymbol{\theta}, \boldsymbol{\phi} \mid \mathcal{D}) \stackrel{c}{=} -D_{\mathrm{KL}}\left(q_{\mathcal{D}, \phi}(\boldsymbol{z}) \| p_{\boldsymbol{\theta}}(\boldsymbol{z})\right) - \mathbb{E}_{\mathcal{D}_{\mathcal{D}, \phi}(\boldsymbol{z})}\left[D_{\mathrm{KL}}\left(q_{\phi}(\boldsymbol{x} \mid \boldsymbol{z}) \| p_{\boldsymbol{\theta}}(\boldsymbol{x} \mid \boldsymbol{z})\right)\right]$$

- $D_{\mathrm{KL}}\left(q_{\phi}(\boldsymbol{z}) \| p_{\boldsymbol{\theta}}(\boldsymbol{z})\right)$: Difference between the inference marginal and generative prior
- $D_{\mathrm{KL}}\left(q_{\phi}(\boldsymbol{x} \mid \boldsymbol{z}) \| p_{\boldsymbol{\theta}}(\boldsymbol{x} \mid \boldsymbol{z})\right)$: Reconstruction error

$\rightarrow \boldsymbol{x}$ is typically of much higher dimensionality than $\boldsymbol{z}$, the reconstruction error term usually dominates. Consequently, if there is a conflict between these two objectives (e.g., due to limited modeling power), the VAE will favor reconstruction accuracy over posterior inference.

Visual Illustration

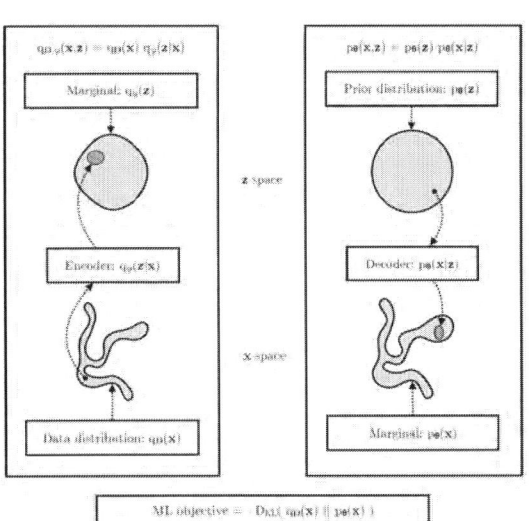

Issue with Basic VAE

VAEs often generate somewhat blurry images.

Consider the common case where the decoder is a Gaussian with fixed variance:

$$\log p_{\boldsymbol{\theta}}(\boldsymbol{x} \mid \boldsymbol{z}) = -\frac{1}{2\sigma^2} \|\boldsymbol{x} - d_{\boldsymbol{\theta}}(\boldsymbol{z})\|_2^2 + \text{const}$$

Here, $d_{\boldsymbol{\theta}}$ is the decoder function and σ is the fixed variance.

Issue with Basic VAE (Cont'd)

Let $e_\phi(\boldsymbol{x}) = \mathbb{E}\left[q_\phi(\boldsymbol{z} \mid \boldsymbol{x})\right]$ be the encoding of $\boldsymbol{x}$, and
$\mathcal{X}(\boldsymbol{z}) = \{\boldsymbol{x} : e_\phi(\boldsymbol{x}) = \boldsymbol{z}\}$ be the set of inputs that get mapped to $\boldsymbol{z}$.

For a fixed inference network, when using squared reconstruction loss, the optimal setting of the generator parameters is:

$$d_{\boldsymbol{\theta}}(\boldsymbol{z}) = \mathbb{E}[\boldsymbol{x} : \boldsymbol{x} \in \mathcal{X}(\boldsymbol{z})]$$

Thus the decoder should predict the average of all inputs $\boldsymbol{x}$ that map to that $\boldsymbol{z}$, resulting in blurry images.

The β-VAE Objective

To resolve this, the β-VAE objective is introduced:

$$\mathcal{L}_\beta(\boldsymbol{\theta}, \boldsymbol{\phi} \mid \boldsymbol{x}) = \underbrace{-\mathbb{E}_{q_\phi(\boldsymbol{z}|\boldsymbol{x})}\left[\log p_{\boldsymbol{\theta}}(\boldsymbol{x} \mid \boldsymbol{z})\right]}_{\mathcal{L}_E} + \beta \underbrace{D_{\mathrm{KL}}\left(q_\phi(\boldsymbol{z} \mid \boldsymbol{x}) \| p_{\boldsymbol{\theta}}(\boldsymbol{z})\right)}_{\mathcal{L}_R},$$

where $\mathcal{L}_E$ is the reconstruction error (negative log likelihood), and $\mathcal{L}_R$ is the KL regularizer.

Interpretation of β in β-VAE

- When $\beta = 1$, we recover the standard VAE objective.
- When $\beta = 0$, we recover the standard autoencoder objective.

By varying β between 0 and infinity, we can make different trade-offs between reconstruction error and the information stored in the latents.

- For $\beta < 1$: Storing more bits about each input can reduce the blurriness in reconstructed images.
- For $\beta > 1$: The representation becomes more compressed.

Advantages of Using $\beta > 1$

- Advantage of $\beta > 1$: Encourages the learning of a "disentangled" latent representation where each dimension represents a different factor of variation in the input.
- Total Correlation (TC): Defined as

$$\mathrm{TC}(\boldsymbol{z}) = \sum_k \mathbb{H}(z_k) - \mathbb{H}(\boldsymbol{z}) = D_{\mathrm{KL}}\left(p(\boldsymbol{z}) \| \prod_k p_k(z_k)\right)$$

TC is zero if the components of $\boldsymbol{z}$ are mutually independent, indicating a disentangled representation.
- Effect of $\beta > 1$: Reduces TC, leading to a less disentangled latent representation.

Limitations of Using $\beta > 1$

- Nonlinearity and Identifiability: Nonlinear latent variable models are unidentifiable. There can be an equivalent fully entangled representation with the same likelihood.
- Recovery of Latent Representation: Adjusting β alone isn't sufficient. Proper inductive bias through the encoder, decoder, prior, dataset, or learning algorithm is necessary.

InfoVAE Objective

In the standard VAE model, the ELBO objective has certain limitations:

- Tendency to ignore the latent code when the decoder is powerful
- Tendency to learn a poor posterior approximation due to the mismatch between the KL terms in data space and latent space

The InfoVAE objective tries to address these issues by introducing a generalized form of the objective:

$$\mathcal{L}(\boldsymbol{\theta}, \boldsymbol{\phi} \mid \boldsymbol{x}) = -\lambda D_{\mathrm{KL}}(q_\phi(\boldsymbol{z}) \| p_\theta(\boldsymbol{z})) - \mathbb{E}_{q_\phi(\boldsymbol{z})}\left[D_{\mathrm{KL}}(q_\phi(\boldsymbol{x} \mid \boldsymbol{z}) \| p_\theta(\boldsymbol{x} \mid \boldsymbol{z}))\right] + \alpha \mathbb{I}_q(\boldsymbol{x}; \boldsymbol{z})$$

Reminder: Mutual Information

The mutual information between random variables X and Y is defined as:

$$\mathbb{I}(X;Y) \triangleq D_{\mathrm{KL}}(p(x,y) \| p(x)p(y)) = \sum_{y \in Y} \sum_{x \in X} p(x,y) \log \frac{p(x,y)}{p(x)p(y)}$$

MI measures the information gain if we update from a model $p(x)p(y)$ to $p(x,y)$.

It can also be re-expressed in terms of entropies:

$$\begin{aligned}\mathbb{I}(X;Y) &= \mathbb{H}(X) - \mathbb{H}(X|Y) \\ &= \mathbb{H}(Y) - \mathbb{H}(Y|X)\end{aligned}$$

$\to$ MI between X and Y: the reduction in uncertainty about X after observing Y, or the reduction in uncertainty about Y after observing X.

Intractability of Mutual Information

The mutual information term, $\mathbb{I}_q(x; z)$, makes the objective intractable:

$$\mathbb{I}_q(x; z) = -\mathbb{E}_{q_\phi(x,z)} \left[\log \frac{q_\phi(z)}{q_\phi(z \mid x)} \right]$$

However, the objective can be rewritten.

Rewriting the Objective

Using certain manipulations, the InfoVAE objective can be rewritten in a more manageable form. The details are shown below:

$$\mathcal{L} = \mathbb{E}_{p_\mathcal{D}(x)} \left[\mathbb{E}_{q_\phi(z\mid x)} [\log p_\theta(x \mid z)] \right] - (1-\alpha)\mathbb{E}_{p_\mathcal{D}(x)} [D_{\text{KL}}(q_\phi(z \mid x) \| p_\theta(z))]$$
$$-(\alpha + \lambda - 1)D_{\text{KL}}(q_\phi(z) \| p_\theta(z)) - \mathbb{E}_{p_\mathcal{D}(x)}[\log p_\mathcal{D}(x)]$$

→ The last term is a constant we can ignore. The first two terms can be optimized using the methods for fitting standard VAE. The third term can be optimized by sampling $x \sim p_\mathcal{D}(x)$ and $z \sim q_\phi(z \mid x)$.

Optimality Theorem

Under certain conditions, the approximate InfoVAE loss can be globally optimized. The theorem is as follows:

> **Theorem 18**
> Let $\mathcal{X}$ and $\mathcal{Z}$ be continuous spaces, and $\alpha < 1$ (to bound the MI) and $\lambda > 0$. For any fixed value of $\mathbb{I}_q(x; z)$, the approximate InfoVAE loss is globally optimized if $p_\theta(x) = p_D(x)$ and $q_\phi(z \mid x) = p_\theta(z \mid x)$.

Connection with β-VAEs

If we set $\alpha = 0$ and $\lambda = 1$, the objective reduces to the original ELBO. For freely chosen $\lambda > 0$ but with $\alpha = 1 - \lambda$, we get the β-VAE.

Multimodal VAEs

- VAEs can be extended to create joint distributions over multiple types of variables, like images and text.
- This extended version is commonly known as a Multimodal VAE, or MVAE.

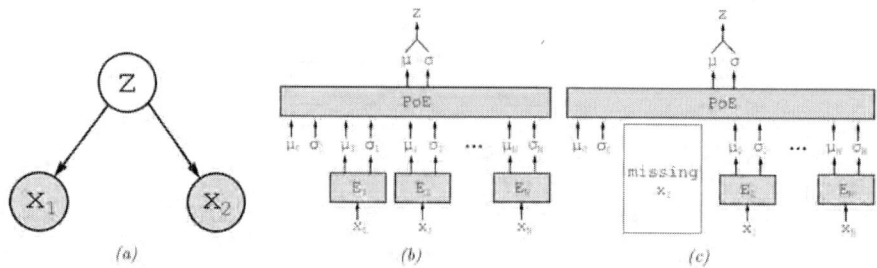

Figure: Illustration of multi-modal VAE.

Mathematical Formulation of MVAE

Let us assume there are M modalities. We assume they are conditionally independent given the latent code, and hence the generative model has the form:

$$p_\theta(x_1, \ldots, x_M, z) = p(z) \prod_{m=1}^{M} p_\theta(x_m \mid z)$$

Then the standard ELBO is given by:

$$\mathcal{L}(\theta, \phi \mid \mathbf{X}) = \mathbb{E}_{q_\phi(z \mid \mathbf{X})} \left[\sum_m \log p_\theta(x_m \mid z) \right] - D_{\mathrm{KL}}(q_\phi(z \mid \mathbf{X}) \| p(z)),$$

where $\mathbf{X} = (x_1, \ldots, x_M)$ is the observed data.

Weighted ELBO

The different likelihood terms $p(x_m \mid z)$ may have different dynamic ranges (e.g., Gaussian pdf for pixels, and categorical pmf for text)

Introducing weight terms $\lambda_m \geq 0$ for each likelihood and $\beta \geq 0$ for KL regularization, we get:

$$\mathcal{L}(\theta, \phi \mid \mathbf{X}) = \mathbb{E}_{q_\phi(z \mid \mathbf{X})} \left[\sum_m \lambda_m \log p_\theta(x_m \mid z) \right] - \beta D_{\mathrm{KL}}(q_\phi(z \mid \mathbf{X}) \| p(z))$$

Unpaired Data

- Data Alignment Issue: Often, we lack sufficient paired data across all M modalities.
- There may be ample data for images (modality 1) and text (modality 2), but very few paired (image, text) examples.
- Loss Function: It becomes crucial to generalize the loss function to handle this scarcity of paired data. Let $O_m = 1$ if modality m is observed, and $O_m = 0$ if it is missing. The objective function is given by:

$$\mathcal{L}(\theta, \phi \mid \mathbf{X}) = \mathbb{E}_{q_\phi(z \mid \mathbf{X})} \left[\sum_{m : O_m = 1} \lambda_m \log p_\theta(x_m \mid z) \right] - \beta D_{\mathrm{KL}}(q_\phi(z \mid \mathbf{X}) \| p(z)),$$

where $\mathbf{X} = \{x_m : O_m = 1\}$ is the visible features.

Key Problem

The challenge is to compute the posterior $q_\phi(z \mid \mathbf{X})$ given different subsets of features.

- How to compute $q_\phi(z \mid x_1)$ if only an image is given?
- What about $q_\phi(z \mid x_2)$ if only text is given?

This issue is generally relevant in VAE with missing inputs. We can compute the optimal form for $q_\phi(z \mid \mathbf{X})$ given set of inputs by computing the exact posterior under the model, which is given by

$$p(z \mid \mathbf{X}) = \frac{p(z)p(x_1,\ldots,x_M \mid z)}{p(x_1,\ldots,x_M)} = \frac{p(z)}{p(x_1,\ldots,x_M)} \prod_{m=1}^{M} p(x_m \mid z)$$

$$= \frac{p(z)}{p(x_1,\ldots,x_M)} \prod_{m=1}^{M} \frac{p(z \mid x_m)p(x_m)}{p(z)}$$

$$\propto p(z) \prod_{m=1}^{M} \frac{p(z \mid x_m)}{p(z)} \approx p(z) \prod_{m=1}^{M} \tilde{q}(z \mid x_m)$$

Computing the Posterior

If we use Gaussian distributions for the prior $p(z) = \mathcal{N}\left(z \mid \mu_0, \Lambda_0^{-1}\right)$ and marginal posterior ratio $\tilde{q}(z \mid x_m) = \mathcal{N}\left(z \mid \mu_m, \Lambda_m^{-1}\right)$, then we can compute the product of Gaussians:

$$\prod_{m=0}^{M} \mathcal{N}\left(z \mid \mu_m, \Lambda_m^{-1}\right) \propto \mathcal{N}(z \mid \mu, \Sigma),$$

$$\Sigma = \left(\sum_m \Lambda_m\right)^{-1},$$

$$\mu = \Sigma \left(\sum_m \Lambda_m \mu_m\right)$$

Training the Model

We need to train individual expert recognition models $q(z \mid x_m)$ and the joint model $q(z \mid \mathbf{X})$.

$$\mathsf{L}(\boldsymbol{\theta},\boldsymbol{\phi} \mid \mathbf{X}) = \mathsf{L}(\boldsymbol{\theta},\boldsymbol{\phi} \mid x_1, \ldots, x_M) + \sum_{m=1}^{M} \mathsf{L}(\boldsymbol{\theta},\boldsymbol{\phi} \mid x_m) + \sum_{j \in \mathcal{J}} \mathsf{L}(\boldsymbol{\theta},\boldsymbol{\phi} \mid \mathbf{X}_j)$$

Semisupervised VAEs

Consider extending VAEs to the semi-supervised learning setting where we have both:

- Labeled data: $\mathcal{D}_L = \{(x_n, y_n)\}$
- Unlabeled data: $\mathcal{D}_U = \{(x_n)\}$

We focus on the M2 model.

Generative Model

The generative model has the following form:

$$p_\theta(x, y) = p_\theta(y)p_\theta(x \mid y)$$
$$= p_\theta(y) \int p_\theta(x \mid y, z) p_\theta(z) dz$$

- z is a latent variable
- $p_\theta(z) = \mathcal{N}(z \mid 0, \mathbf{I})$ is the latent prior
- $p_\theta(y) = \text{Cat}(y \mid \pi)$ is the label prior

Likelihood

The likelihood $p_\theta(x \mid y, z)$ is a function, often Gaussian, parameterized by a deep neural network f:

$$p_\theta(x \mid y, z) = p(x \mid f_\theta(y, z))$$

Key idea: Data is generated according to both a latent class variable y and the continuous latent variable z. y is observed for labeled data and unobserved for unlabeled data.

Variants of VAE | Semi-Supervised VAEs: M2 Model
Likelihood for Labeled Data $\log p_\theta(x, y)$

To compute the likelihood for the labeled data $\log p_\theta(x, y)$, we need to marginalize over z, which we use an inference network of the form:

$$q_\phi(z \mid y, x) = \mathcal{N}\left(z \mid \mu_\phi(y, x), \operatorname{diag}\left(\sigma_\phi(y, x)\right)\right)$$

The variational lower bound is given by:

$$\log p_\theta(x, y) \geq \mathbb{E}_{q_\phi(z \mid x, y)}\left[\log p_\theta(x \mid y, z) + \log p_\theta(y) + \log p_\theta(z) - \log q_\phi(z \mid x, y)\right]$$
$$= -\mathcal{L}(x, y)$$

This is standard for VAEs. The only difference is the observation of two kinds of data: x and y.

Variants of VAE | Semi-Supervised VAEs: M2 Model
Likelihood for Unlabeled Data $\log p_\theta(x)$

To compute the likelihood for the unlabeled data $p_\theta(x)$, we need to marginalize over z and y. We achieve this using an inference network of the form:

$$q_\phi(z, y \mid x) = q_\phi(z \mid x) q_\phi(y \mid x)$$
$$q_\phi(z \mid x) = \mathcal{N}\left(z \mid \mu_\phi(x), \operatorname{diag}\left(\sigma_\phi(x)\right)\right)$$
$$q_\phi(y \mid x) = \operatorname{Cat}\left(y \mid \pi_\phi(x)\right)$$

The variational lower bound is:

$$\log p_\theta(x) \geq \mathbb{E}_{q_\phi(z, y \mid x)}\left[\log p_\theta(x \mid y, z) + \log p_\theta(y) + \log p_\theta(z) - \log q_\phi(z, y \mid x)\right]$$
$$= -\sum_y q_\phi(y \mid x) \mathcal{L}(x, y) + \mathbb{H}\left(q_\phi(y \mid x)\right)$$
$$= -\mathcal{U}(x)$$

Objective Function

The discriminative classifier $q_\phi(y \mid x)$ is used only to compute the log-likelihood of the unlabeled data, which is not ideal. Thus, we add an extra classification loss on the supervised data to form the overall objective function:

$$\mathcal{L}(\boldsymbol{\theta}) = \mathbb{E}_{(x,y)\sim \mathcal{D}_L}[\mathcal{L}(x,y)] + \mathbb{E}_{x\sim \mathcal{D}_U}[\mathcal{U}(x)] + \alpha \mathbb{E}_{(x,y)\sim \mathcal{D}_L}[-\log q_\phi(y \mid x)]$$

Here, $\mathcal{D}_L$ is the labeled data, $\mathcal{D}_U$ is the unlabeled data, and α is a hyperparameter.

Introduction to Hierarchical VAEs

- A Hierarchical VAE (HVAE) contains L stochastic layers.
- Generative model is defined as:

$$p_\theta(x, z_{1:L}) = p_\theta(z_L) \left[\prod_{l=L-1}^{1} p_\theta(z_l \mid z_{l+1}) \right] p_\theta(x \mid z_1)$$

Non-Markovian Hierarchical VAEs

- Each z_l depends on all higher-level stochastic variables $z_{l+1:L}$.

$$p_\theta(x, z) = p_\theta(z_L) \left[\prod_{l=L-1}^{1} p_\theta(z_l \mid z_{l+1:L}) \right] p_\theta(x \mid z_{1:L})$$

Generative Model with Skip Connections

- The likelihood is now $p_\theta(x \mid z_{1:L})$.
- Analogous to adding skip connections.

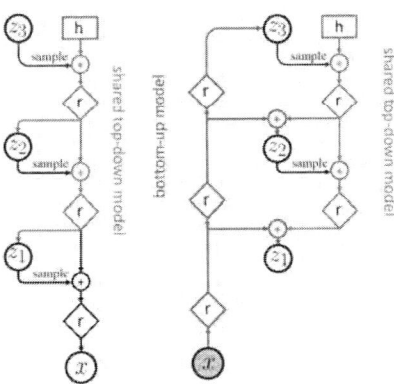

Figure: Hierarchical VAEs with 3 stochastic layers.

Bottom-up vs top-down inference: Models

To perform inference in a hierarchical VAE, we could use a bottom-up inference model:

$$q_\phi(\boldsymbol{z} \mid \boldsymbol{x}) = q_\phi(\boldsymbol{z}_1 \mid \boldsymbol{x}) \prod_{l=2}^{L} q_\phi(\boldsymbol{z}_l \mid \boldsymbol{x}, \boldsymbol{z}_{1:l-1})$$

A better approach is to use a top-down inference model:

$$q_\phi(\boldsymbol{z} \mid \boldsymbol{x}) = q_\phi(\boldsymbol{z}_L \mid \boldsymbol{x}) \prod_{l=L-1}^{1} q_\phi(\boldsymbol{z}_l \mid \boldsymbol{x}, \boldsymbol{z}_{l+1:L})$$

Reason for Top-down Inference Model

The reason the top-down inference model is better is that it more closely approximates the true posterior:

$$p_\theta(\boldsymbol{z}_l \mid \boldsymbol{x}, \boldsymbol{z}_{l+1:L}) \propto p_\theta(\boldsymbol{z}_l \mid \boldsymbol{z}_{l+1:L}) p_\theta(\boldsymbol{x} \mid \boldsymbol{z}_l, \boldsymbol{z}_{l+1:L})$$

Thus the posterior combines the top-down prior term $p_\theta(\boldsymbol{z}_l \mid \boldsymbol{z}_{l+1:L})$ with the bottom-up likelihood term $p_\theta(\boldsymbol{x} \mid \boldsymbol{z}_l, \boldsymbol{z}_{l+1:L})$.

Gaussian Approximation

We can approximate this posterior by defining:

$$q_\phi(z_l \mid x, z_{l+1:L}) \propto p_\theta(z_l \mid z_{l+1:L}) \tilde{q}_\phi(z_l \mid x, z_{l+1:L})$$

where $\tilde{q}_\phi(z_l \mid x, z_{l+1:L})$ is a learned Gaussian approximation to the bottom-up likelihood.

ELBO Formulation

The ELBO can be written as:

$$\mathcal{L}(\theta, \phi \mid x) = \mathbb{E}_{q_\phi(z\mid x)}[\log p_\theta(x \mid z)] - D_{\mathrm{KL}}(q_\phi(z_L \mid x) \| p_\theta(z_L))$$
$$- \sum_{l=L-1}^{1} \mathbb{E}_{q_\phi(z_{>l}\mid x)}[D_{\mathrm{KL}}(q_\phi(z_l \mid x, z_{>l}) \| p_\theta(z_l \mid z_{>l}))]$$

where

$$q_\phi(z_{>l} \mid x) = \prod_{i=l+1}^{L} q_\phi(z_i \mid x, z_{>i})$$

is the approximate posterior above layer l (i.e., the parents of z_l).

Deep VAE

- There have been many papers exploring different kinds of HVAE models. "Very deep VAE" is simple but yields state-of-the-art results.
- The architecture is a simple convolutional VAE with bidirectional inference.
    - For each layer, the prior and posterior are diagonal Gaussians.
    - Nearest-neighbor upsampling worked better than transposed convolution.

Architecture

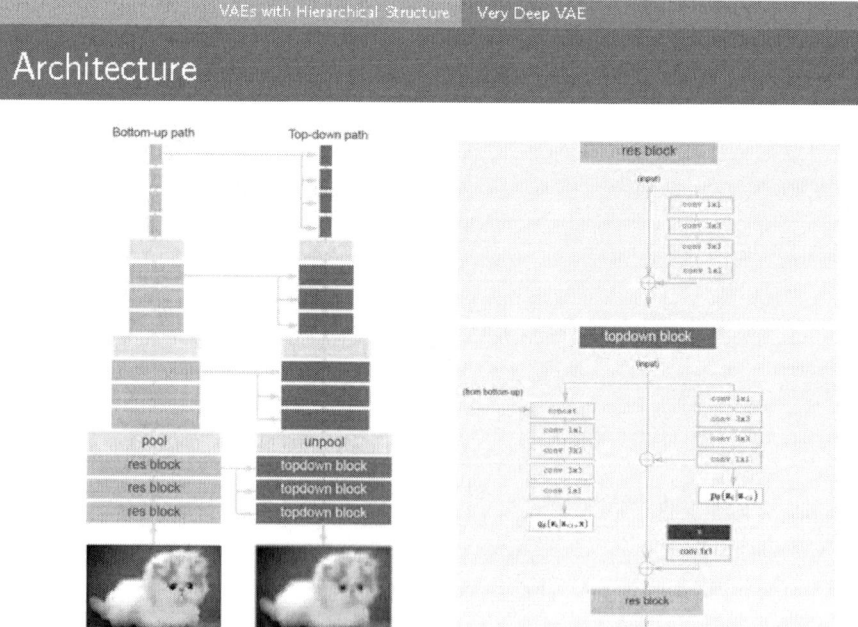

Figure: The top-down encoder used by the hierarchical VAE

Benefits

- Low-resolution latents capture a lot of the global structure.
- High-resolution latents fill in details, making the image look more realistic.

This suggests the model could be useful for lossy compression.

Unconditional Sampling

The model can also be used for unconditional sampling at multiple resolutions.

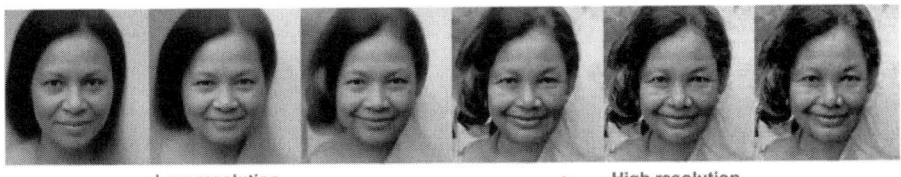

Low resolution ⟶ High resolution

Figure: Samples from a VDVAE model trained on FFHQ dataset.

Chapter 9.
Diffusion Model

Introduction

VAEs with Hierarchical Structure   Very Deep VAE

Given observed samples x from a distribution of interest, the goal of a generative model is to learn to model its true data distribution $p(x)$.

In this section we explore and review diffusion models.

Background: ELBO, VAE, and Hierarchical VAE

For many modalities, the data can be thought of as generated by an unseen latent variable z.

Latent Variables

- Latent Variables: Data observed can be generated by unseen latent variables, denoted by z.
- Higher-level Representations: Objects in the real world may be generated by higher-level properties like color, size, shape, etc.
  - Projection: Observations can be considered as lower-dimensional instantiations of these abstract concepts.
- Inference: Unobservable latent variables can be approximated to describe observed data.
- Dimensionality: Often seek lower-dimensional latent spaces.
- Why Lower Dimensionality: Lower dimensions can act as data compression and uncover meaningful structure.

Evidence Lower Bound (ELBO)

Mathematically, we consider a joint distribution $p(x, z)$ to model latent variables and observed data. The goal in likelihood-based generative modeling is to maximize the likelihood $p(x)$ for all observed x.

Two approaches to recover $p(x)$:

$$p(x) = \int p(x, z) dz \qquad (11)$$

or using the chain rule of probability:

$$p(x) = \frac{p(x, z)}{p(z \mid x)} \qquad (12)$$

Issues with Direct Computation

Direct computation of $p(x)$ is challenging:

- Equation 11 involves intractable integration.
- Equation 12 requires a ground truth latent encoder $p(z \mid x)$.

Evidence Lower Bound (ELBO)

We derive the ELBO as a lower bound of the log likelihood of observed data:
$$\mathbb{E}_{q_\phi(z|x)} \left[\log \frac{p(x, z)}{q_\phi(z \mid x)} \right]$$

Explicit relationship with evidence:
$$\log p(x) \geq \mathbb{E}_{q_\phi(z|x)} \left[\log \frac{p(x, z)}{q_\phi(z \mid x)} \right]$$

Variational Distribution q_ϕ

Here, $q_\phi(z \mid x)$ is an approximate variational distribution with parameters ϕ that we optimize. It estimates the true posterior $p(z \mid x)$.

Why Maximize ELBO?

Maximizing the ELBO gives us components to model the true data distribution and sample from it, effectively learning a generative model. It serves as a proxy objective to optimize a latent variable model.

Derivation using Equation 11

Let us begin by deriving the ELBO using Equation 11:

$$\log p(\boldsymbol{x}) = \log \int p(\boldsymbol{x}, \boldsymbol{z}) d\boldsymbol{z}$$
$$= \log \mathbb{E}_{q_\phi(\boldsymbol{z}|\boldsymbol{x})} \left[\frac{p(\boldsymbol{x}, \boldsymbol{z})}{q_\phi(\boldsymbol{z} \mid \boldsymbol{x})} \right]$$
$$\geq \mathbb{E}_{q_\phi(\boldsymbol{z}|\boldsymbol{x})} \left[\log \frac{p(\boldsymbol{x}, \boldsymbol{z})}{q_\phi(\boldsymbol{z} \mid \boldsymbol{x})} \right] \quad \text{(Jensen's Inequality)}$$

Limitations of First Derivation

However, this does not supply much useful information. Crucially, Jensen's Inequality handwaves away why the ELBO is a lower bound. To better understand, let's derive it again using Equation 12.

Derivation using Equation 12

$$\log p(x)$$
$$= \log p(x) \int q_\phi(z \mid x) dz \qquad \text{(Multiply by } 1 = \int q_\phi(z \mid x) dz\text{)}$$
$$= \int q_\phi(z \mid x)(\log p(x)) dz \qquad \text{(Bring evidence into integral)}$$
$$= \mathbb{E}_{q_\phi(z \mid x)}[\log p(x)] \qquad \text{(Definition of Expectation)}$$
$$= \mathbb{E}_{q_\phi(z \mid x)}\left[\log \frac{p(x, z)}{p(z \mid x)}\right] \qquad \text{(Apply Equation 12)}$$
$$= \mathbb{E}_{q_\phi(z \mid x)}\left[\log \frac{p(x, z) q_\phi(z \mid x)}{p(z \mid x) q_\phi(z \mid x)}\right] \qquad \text{(Multiply by } 1 = \frac{q_\phi(z \mid x)}{q_\phi(z \mid x)}\text{)}$$
$$= \mathbb{E}_{q_\phi(z \mid x)}\left[\log \frac{p(x, z)}{q_\phi(z \mid x)}\right] + \mathbb{E}_{q_\phi(z \mid x)}\left[\log \frac{q_\phi(z \mid x)}{p(z \mid x)}\right] \qquad \text{(Split the Expectation)}$$
$$= \mathbb{E}_{q_\phi(z \mid x)}\left[\log \frac{p(x, z)}{q_\phi(z \mid x)}\right] + D_{\mathrm{KL}}\left(q_\phi(z \mid x) \parallel p(z \mid x)\right) \qquad \text{(Definition of KL Divergence)}$$
$$\geq \mathbb{E}_{q_\phi(z \mid x)}\left[\log \frac{p(x, z)}{q_\phi(z \mid x)}\right] \qquad \text{(KL Divergence always } \geq 0\text{)}$$

Insights from Second Derivation

From this derivation, we observe that the evidence is the ELBO plus the KL Divergence between the approximate and true posterior.
Understanding this term is key to understanding the relationship between the ELBO and the evidence, as well as why optimizing the ELBO is an appropriate objective.

Why ELBO is a Lower Bound

The difference between the evidence and the ELBO is a strictly non-negative KL term, thus the value of the ELBO can never exceed the evidence:

$$\text{Evidence} - \text{ELBO} = D_{\text{KL}}\left(q_\phi(z \mid x) \parallel p(z \mid x)\right) \geq 0$$

Why Maximize ELBO?

- Goal: Learn the latent structure described by z to optimize the variational posterior $q_\phi(z \mid x)$ to match the true posterior $p(z \mid x)$.
    - Method: Minimize KL Divergence between the variational and true posterior.
    - Challenge: Intractable to minimize KL Divergence directly due to lack of access to $p(z \mid x)$.
- ELBO Maximization:
    - Reason: Data likelihood $\log p(x)$ is constant with respect to ϕ.
    - Result: Maximizing ELBO minimizes KL Divergence as they sum up to a constant.
- Utility of ELBO:
    - Proxy: Maximizing ELBO helps model the true latent posterior effectively.
    - Post-Training: Can be used to estimate likelihood of observed or generated data.

Variational Autoencoders I

- VAE Objective: Maximize the ELBO using variational approach.
  - Optimize $q_\phi(z \mid x)$ within a family of distributions parameterized by ϕ.
  - Resemblance to traditional autoencoders through bottlenecking representation.

Variational Autoencoders II

- ELBO Term Dissection:

$$\mathbb{E}_{q_\phi(z \mid x)} \left[\log \frac{p(x, z)}{q_\phi(z \mid x)} \right]$$

$$= \mathbb{E}_{q_\phi(z \mid x)} \left[\log \frac{p_\theta(x \mid z) p(z)}{q_\phi(z \mid x)} \right] \quad \text{(Chain Rule of Probability)}$$

$$= \mathbb{E}_{q_\phi(z \mid x)} \left[\log p_\theta(x \mid z) \right] + \mathbb{E}_{q_\phi(z \mid x)} \left[\log \frac{p(z)}{q_\phi(z \mid x)} \right] \quad \text{(Split the Expectation)}$$

$$= \underbrace{\mathbb{E}_{q_\phi(z \mid x)} \left[\log p_\theta(x \mid z) \right]}_{\text{reconstruction term}} - \underbrace{D_{\mathrm{KL}} \left(q_\phi(z \mid x) \parallel p(z) \right)}_{\text{prior matching term}} \quad \text{(Definition of KL Divergence)}$$

Variational Autoencoders III

- Components:
  - Encoder: $q_\phi(z \mid x)$, transforms inputs to a distribution over latents.
  - Decoder: $p_\theta(x \mid z)$, converts latent vector z to an observation x.
- Intuitive Descriptions:
  - Reconstruction Term: Measures reconstruction likelihood of the decoder, ensuring effective latents for data regeneration.
  - Prior Matching Term: Measures similarity to a prior over latents, preventing encoder collapse into a Dirac delta function.

Graphical Representation of a Variational Autoencoder

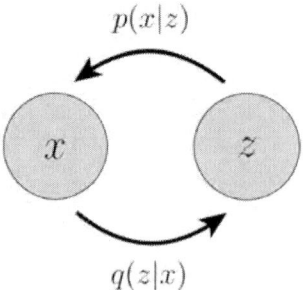

Figure: A Variational Autoencoder graphically represented.

Joint optimization over parameters ϕ and θ I

- ELBO Optimization: Maximize first term (reconstruction) and minimize second term (KL Divergence).
- Encoder and Prior Distributions:

$$q_\phi(z \mid x) = \mathcal{N}\left(z; \boldsymbol{\mu}_\phi(x), \sigma_\phi^2(x)\mathbf{I}\right),$$
$$p(z) = \mathcal{N}(z; 0, \mathbf{I}).$$

Joint optimization over parameters ϕ and θ II

- Objective Function:

$$\arg\max_{\phi,\theta} \mathbb{E}_{q_\phi(z|x)}\left[\log p_\theta(x \mid z)\right] - D_{\mathrm{KL}}\left(q_\phi(z \mid x) \| p(z)\right).$$

- Monte Carlo Estimation:

$$\approx \arg\max_{\phi,\theta} \sum_{l=1}^{L} \log p_\theta\left(x \mid z^{(l)}\right) - D_{\mathrm{KL}}\left(q_\phi(z \mid x) \| p(z)\right),$$

where $z^{(l)}$ are sampled from $q_\phi(z \mid x)$.
- Non-Differentiability Issue: Loss is computed on $z^{(l)}$ generated by a non-differentiable stochastic sampling procedure.

Reparameterization Trick I

- Reparameterization Trick: Addresses non-differentiability in VAEs, applicable when $q_\phi(z \mid x)$ models certain distributions like the multivariate Gaussian.
- Random Variable as Deterministic Function:

$$x = \mu + \sigma\epsilon \quad \text{with} \quad \epsilon \sim \mathcal{N}(\epsilon; 0, 1)$$

- Utility in Gaussian Distributions:
  - Shift mean from zero to target μ by addition.
  - Stretch variance to target σ^2.

Reparameterization Trick II

- VAE Reparameterization:

$$z = \boldsymbol{\mu}_\phi(\boldsymbol{x}) + \boldsymbol{\sigma}_\phi(\boldsymbol{x}) \odot \epsilon \quad \text{with} \quad \epsilon \sim \mathcal{N}(\epsilon; 0, \mathbf{I})$$

  where $\odot$ is element-wise product.
- Gradient Computability:
  - Gradients can now be computed w.r.t. ϕ.
  - Enables optimization of $\boldsymbol{\mu}_\phi$ and $\boldsymbol{\sigma}_\phi$.

Data Generation and VAEs

After training a VAE, generating new data can be performed by sampling directly from the latent space $p(z)$ and running it through the decoder.

Variational Autoencoders are particularly interesting when the dimensionality of z is less than that of input x.

When a semantically meaningful latent space is learned, latent vectors can be edited before being passed to the decoder to more precisely control the data generated.

Hierarchical Variational Autoencoders I

- Hierarchical Variational Autoencoder (HVAE): Generalizes VAE to multiple hierarchies over latent variables.
- Latent Hierarchy: Latent variables are generated from higher-level, more abstract latents.
- HVAE with T Levels:
    - Each latent can condition on all previous latents.
    - Special Case: Markovian HVAE (MHVAE).
- Markovian HVAE (MHVAE):
    - Generative process is a Markov chain.
    - Each latent z_t conditions only on previous latent z_{t+1}.
- Visual Interpretation: Stacking VAEs on top of each other, also known as a Recursive VAE.

MHVAE: Visual Representation

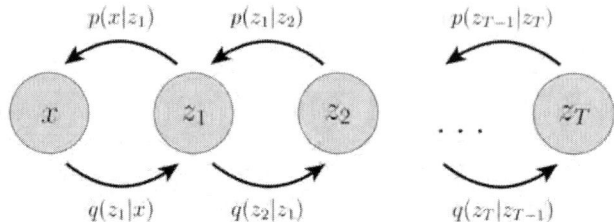

Figure: A Markovian Hierarchical Variational Autoencoder with T hierarchical latents.

Mathematical Formulation I

Focus on a special case: Markovian HVAE (MHVAE).

In MHVAE, the generative process is a Markov chain.

Mathematically, we represent the joint distribution and the posterior as:

$$p\left(\boldsymbol{x}, \boldsymbol{z}_{1:T}\right) = p\left(\boldsymbol{z}_T\right) p_{\boldsymbol{\theta}}\left(\boldsymbol{x} \mid \boldsymbol{z}_1\right) \prod_{t=2}^{T} p_{\boldsymbol{\theta}}\left(\boldsymbol{z}_{t-1} \mid \boldsymbol{z}_t\right)$$

$$q_{\boldsymbol{\phi}}\left(\boldsymbol{z}_{1:T} \mid \boldsymbol{x}\right) = q_{\boldsymbol{\phi}}\left(\boldsymbol{z}_1 \mid \boldsymbol{x}\right) \prod_{t=2}^{T} q_{\boldsymbol{\phi}}\left(\boldsymbol{z}_t \mid \boldsymbol{z}_{t-1}\right)$$

Mathematical Formulation II

Extending the ELBO:

$$\begin{aligned}
\log p(\boldsymbol{x}) &= \log \int p(\boldsymbol{x}, \boldsymbol{z}_{1:T}) \, d\boldsymbol{z}_{1:T} && \text{(Apply Equation 11)} \\
&= \log \int \frac{p(\boldsymbol{x}, \boldsymbol{z}_{1:T}) \, q_\phi(\boldsymbol{z}_{1:T} \mid \boldsymbol{x})}{q_\phi(\boldsymbol{z}_{1:T} \mid \boldsymbol{x})} d\boldsymbol{z}_{1:T} && \left(\text{Multiply by } 1 = \frac{q_\phi(\boldsymbol{z}_{1:T} \mid \boldsymbol{x})}{q_\phi(\boldsymbol{z}_{1:T} \mid \boldsymbol{x})}\right) \\
&= \log \mathbb{E}_{q_\phi(\boldsymbol{z}_{1:T} \mid \boldsymbol{x})} \left[\frac{p(\boldsymbol{x}, \boldsymbol{z}_{1:T})}{q_\phi(\boldsymbol{z}_{1:T} \mid \boldsymbol{x})} \right] && \text{(Definition of Expectation)} \\
&\geq \mathbb{E}_{q_\phi(\boldsymbol{z}_{1:T} \mid \boldsymbol{x})} \left[\log \frac{p(\boldsymbol{x}, \boldsymbol{z}_{1:T})}{q_\phi(\boldsymbol{z}_{1:T} \mid \boldsymbol{x})} \right] && \text{(Apply Jensen's Inequality)}
\end{aligned}$$

Mathematical Formulation III

Alternate form after plugging in joint distribution and posterior:

$$\mathbb{E}_{q_\phi(\boldsymbol{z}_{1:T} \mid \boldsymbol{x})} \left[\log \frac{p(\boldsymbol{x}, \boldsymbol{z}_{1:T})}{q_\phi(\boldsymbol{z}_{1:T} \mid \boldsymbol{x})} \right] = \mathbb{E}_{q_\phi(\boldsymbol{z}_{1:T} \mid \boldsymbol{x})} \left[\log \frac{p(\boldsymbol{z}_T) \, p_\theta(\boldsymbol{x} \mid \boldsymbol{z}_1) \prod_{t=2}^{T} p_\theta(\boldsymbol{z}_{t-1} \mid \boldsymbol{z}_t)}{q_\phi(\boldsymbol{z}_1 \mid \boldsymbol{x}) \prod_{t=2}^{T} q_\phi(\boldsymbol{z}_t \mid \boldsymbol{z}_{t-1})} \right]$$

Variational Diffusion Models: Introduction

A Variational Diffusion Model (VDM) is a Markovian Hierarchical Variational Autoencoder with three key restrictions:

1. Latent dimension equals data dimension
2. Encoder structure is a predefined linear Gaussian model
3. Gaussian parameters of latent encoders vary over time in such a way that the distribution of the latent at final timestep T is a standard Gaussian

Mathematical Formulation I

- Assumptions and Notation:
    - Represent both true data samples and latents as x_t, with $t=0$ for true data and $t \in [1, T]$ for latents.
- VDM Posterior as MHVAE Posterior:

$$q(x_{1:T} \mid x_0) = \prod_{t=1}^{T} q(x_t \mid x_{t-1})$$

Mathematical Formulation II

- Encoder Distribution:
  - Gaussian centered around previous hierarchical latent.
  - Fixed as linear Gaussian model.
  - Encoder transitions:
  $$q(\boldsymbol{x}_t \mid \boldsymbol{x}_{t-1}) = \mathcal{N}(\boldsymbol{x}_t; \sqrt{\alpha_t}\boldsymbol{x}_{t-1}, (1-\alpha_t)\mathbf{I})$$
- Parameter α_t:
  - Potentially learnable.
  - Evolves according to a schedule to ensure $p(\boldsymbol{x}_T)$ is a standard Gaussian.

Mathematical Formulation III

- Joint Distribution of VDM:
$$p(\boldsymbol{x}_{0:T}) = p(\boldsymbol{x}_T) \prod_{t=1}^{T} p_{\boldsymbol{\theta}}(\boldsymbol{x}_{t-1} \mid \boldsymbol{x}_t)$$
$$p(\boldsymbol{x}_T) = \mathcal{N}(\boldsymbol{x}_T; 0, \mathbf{I})$$

- Encoder Distributions:
  - No longer parameterized by ϕ.
  - Only learn conditionals $p_{\boldsymbol{\theta}}(\boldsymbol{x}_{t-1} \mid \boldsymbol{x}_t)$.

Mathematical Formulation IV

- Sampling Procedure:
  - Sample noise from $p(x_T)$.
  - Run denoising transitions for T steps to generate x_0.
- Overall Process:
  - Progressive corruption of an image by adding Gaussian noise until it becomes pure Gaussian noise.

VDM: Visual Representation

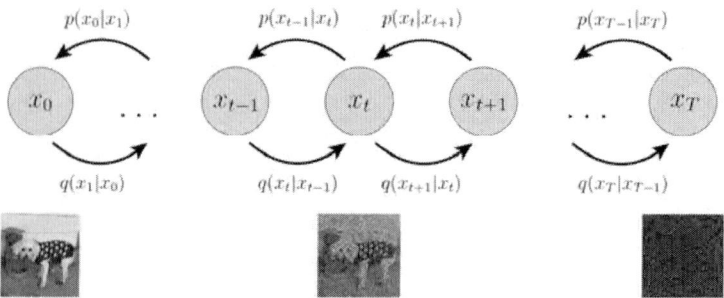

Figure: A visual representation of a VDM.

Variational Diffusion Models

Optimizing VDM by Maximizing ELBO I

$$\log p(x) = \log \int p(x_{0:T}) dx_{1:T}$$

$$= \log \int \frac{p(x_{0:T}) q(x_{1:T} \mid x_0)}{q(x_{1:T} \mid x_0)} dx_{1:T}$$

$$= \log \mathbb{E}_{q(x_{1:T}\mid x_0)} \left[\frac{p(x_{0:T})}{q(x_{1:T} \mid x_0)} \right]$$

$$\geq \mathbb{E}_{q(x_{1:T}\mid x_0)} \left[\log \frac{p(x_{0:T})}{q(x_{1:T} \mid x_0)} \right]$$

$$= \mathbb{E}_{q(x_{1:T}\mid x_0)} \left[\log \frac{p(x_T) \prod_{t=1}^{T} p_\theta (x_{t-1} \mid x_t)}{\prod_{t=1}^{T} q(x_t \mid x_{t-1})} \right]$$

$$= \mathbb{E}_{q(x_{1:T}\mid x_0)} \left[\log \frac{p(x_T) p_\theta(x_0 \mid x_1) \prod_{t=2}^{T} p_\theta(x_{t-1} \mid x_t)}{q(x_T \mid x_{T-1}) \prod_{t=1}^{T-1} q(x_t \mid x_{t-1})} \right]$$

$$= \mathbb{E}_{q(x_{1:T}\mid x_0)} \left[\log \frac{p(x_T) p_\theta(x_0 \mid x_1) \prod_{t=1}^{T-1} p_\theta(x_t \mid x_{t+1})}{q(x_T \mid x_{T-1}) \prod_{t=1}^{T-1} q(x_t \mid x_{t-1})} \right]$$

$$= \mathbb{E}_{q(x_{1:T}\mid x_0)} \left[\log \frac{p(x_T) p_\theta(x_0 \mid x_1)}{q(x_T \mid x_{T-1})} \right] + \mathbb{E}_{q(x_{1:T}\mid x_0)} \left[\log \prod_{t=1}^{T-1} \frac{p_\theta(x_t \mid x_{t+1})}{q(x_t \mid x_{t-1})} \right]$$

Variational Diffusion Models

Optimizing VDM by Maximizing ELBO II

$$= \mathbb{E}_{q(x_{1:T}\mid x_0)} [\log p_\theta(x_0 \mid x_1)] + \mathbb{E}_{q(x_{1:T}\mid x_0)} \left[\log \frac{p(x_T)}{q(x_T \mid x_{T-1})} \right]$$

$$+ \mathbb{E}_{q(x_{1:T}\mid x_0)} \left[\sum_{t=1}^{T-1} \log \frac{p_\theta(x_t \mid x_{t+1})}{q(x_t \mid x_{t-1})} \right]$$

$$= \mathbb{E}_{q(x_{1:T}\mid x_0)} [\log p_\theta(x_0 \mid x_1)] + \mathbb{E}_{q(x_{1:T}\mid x_0)} \left[\log \frac{p(x_T)}{q(x_T \mid x_{T-1})} \right]$$

$$+ \sum_{t=1}^{T-1} \mathbb{E}_{q(x_{1:T}\mid x_0)} \left[\log \frac{p_\theta(x_t \mid x_{t+1})}{q(x_t \mid x_{t-1})} \right]$$

$$= \mathbb{E}_{q(x_1\mid x_0)} [\log p_\theta(x_0 \mid x_1)] + \mathbb{E}_{q(x_{T-1},x_T\mid x_0)} \left[\log \frac{p(x_T)}{q(x_T \mid x_{T-1})} \right]$$

$$+ \sum_{t=1}^{T-1} \mathbb{E}_{q(x_{t-1},x_t,x_{t+1}\mid x_0)} \left[\log \frac{p_\theta(x_t \mid x_{t+1})}{q(x_t \mid x_{t-1})} \right]$$

$$= \underbrace{\mathbb{E}_{q(x_1\mid x_0)} [\log p_\theta(x_0 \mid x_1)]}_{\text{reconstruction term}} - \underbrace{\mathbb{E}_{q(x_{T-1}\mid x_0)} \left[D_{\mathrm{KL}} \left(q(x_T \mid x_{T-1}) \parallel p(x_T) \right) \right]}_{\text{prior matching term}}$$

$$- \sum_{t=1}^{T-1} \underbrace{\mathbb{E}_{q(x_{t-1},x_{t+1}\mid x_0)} \left[D_{\mathrm{KL}} \left(q(x_t \mid x_{t-1}) \parallel p_\theta(x_t \mid x_{t+1}) \right) \right]}_{\text{consistency term}}$$

ELBO Components in VDM I

Individual Components of ELBO:

❶ Reconstruction Term:

$$\mathbb{E}_{q(\boldsymbol{x}_1|\boldsymbol{x}_0)} \left[\log p_{\boldsymbol{\theta}}\left(\boldsymbol{x}_0 \mid \boldsymbol{x}_1\right)\right]$$

- Predicts the log probability of the original data sample given the first-step latent.
- Similar to the term in vanilla VAE and trainable in the same manner.

ELBO Components in VDM II

❷ Prior Matching Term:

$$\mathbb{E}_{q(\boldsymbol{x}_{T-1}|\boldsymbol{x}_0)} \left[D_{\mathrm{KL}}\left(q\left(\boldsymbol{x}_T \mid \boldsymbol{x}_{T-1}\right) \| p\left(\boldsymbol{x}_T\right)\right)\right]$$

- Minimized when the final latent distribution matches the Gaussian prior.
- No trainable parameters and becomes effectively zero for large T.

ELBO Components in VDM III

- Consistency Term:

$$\sum_{t=1}^{T-1} \mathbb{E}_{q(\boldsymbol{x}_{t-1}, \boldsymbol{x}_{t+1}|\boldsymbol{x}_0)} \left[D_{\mathrm{KL}} \left(q\left(\boldsymbol{x}_t \mid \boldsymbol{x}_{t-1}\right) \| p_{\boldsymbol{\theta}} \left(\boldsymbol{x}_t \mid \boldsymbol{x}_{t+1}\right) \right) \right]$$

- Ensures the distribution at $\boldsymbol{x}_t$ is consistent through both forward and backward processes.
- Minimized when $p_{\boldsymbol{\theta}}\left(\boldsymbol{x}_t \mid \boldsymbol{x}_{t+1}\right)$ matches $q\left(\boldsymbol{x}_t \mid \boldsymbol{x}_{t-1}\right)$.
- The cost of optimizing a VDM is primarily dominated by this term, since we must optimize over all timesteps t.

VDM: Visual Representation I

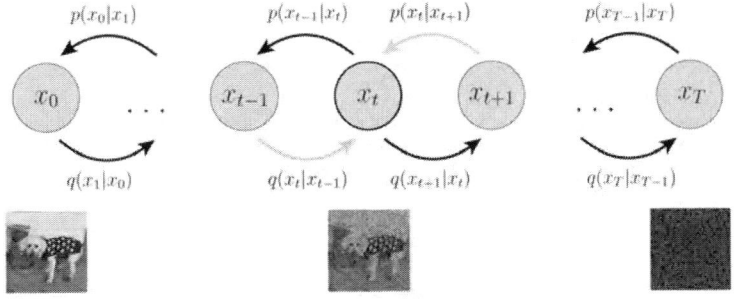

Figure: Under our first derivation, a VDM can be optimized by ensuring that for every intermediate $\boldsymbol{x}_t$, the posterior from the latent above it $p_{\boldsymbol{\theta}}\left(\boldsymbol{x}_t \mid \boldsymbol{x}_{t+1}\right)$ matches the Gaussian corruption of the latent before it $q\left(\boldsymbol{x}_t \mid \boldsymbol{x}_{t-1}\right)$. In this figure, for each intermediate $\boldsymbol{x}_t$, we minimize the difference between the distributions represented by the pink and green arrows.

High Variance Issue I

- Monte Carlo Approximation:
  - All terms of the ELBO are computed as expectations and can be approximated using Monte Carlo estimates.
  - The consistency term is calculated as an expectation over two random variables $\{x_{t-1}, x_{t+1}\}$ at each timestep.
  - This could result in higher variance in its Monte Carlo estimate compared to terms estimated with only one random variable per timestep, especially for large T values..

High Variance Issue II

- Optimizing ELBO with Reduced Variance:
  - Aim to derive a form of the ELBO where each term is an expectation over only one random variable at a time.
  - Key Insight: Rewrite encoder transitions exploiting the Markov property as
  $$q(x_t \mid x_{t-1}) = q(x_t \mid x_{t-1}, x_0)$$
  - The extra conditioning term x_0 becomes superfluous due to the Markov property.
  - Then, according to Bayes rule, we can rewrite each transition as:
  $$q(x_t \mid x_{t-1}, x_0) = \frac{q(x_{t-1} \mid x_t, x_0) \, q(x_t \mid x_0)}{q(x_{t-1} \mid x_0)}$$

Derivation of a Low-Variance ELBO I

$$\log p(x) \geq \mathbb{E}_{q(x_{1:T}|x_0)} \left[\log \frac{p(x_{0:T})}{q(x_{1:T}|x_0)} \right]$$

$$= \mathbb{E}_{q(x_{1:T}|x_0)} \left[\log \frac{p(x_T) \prod_{t=1}^{T} p_\theta(x_{t-1}|x_t)}{\prod_{t=1}^{T} q(x_t|x_{t-1})} \right]$$

$$= \mathbb{E}_{q(x_{1:T}|x_0)} \left[\log \frac{p(x_T) p_\theta(x_0|x_1) \prod_{t=2}^{T} p_\theta(x_{t-1}|x_t)}{q(x_1|x_0) \prod_{t=2}^{T} q(x_t|x_{t-1})} \right]$$

$$= \mathbb{E}_{q(x_{1:T}|x_0)} \left[\log \frac{p(x_T) p_\theta(x_0|x_1) \prod_{t=2}^{T} p_\theta(x_{t-1}|x_t)}{q(x_1|x_0) \prod_{t=2}^{T} q(x_t|x_{t-1}, x_0)} \right]$$

$$= \mathbb{E}_{q(x_{1:T}|x_0)} \left[\log \frac{p_\theta(x_T) p_\theta(x_0|x_1)}{q(x_1|x_0)} + \log \prod_{t=2}^{T} \frac{p_\theta(x_{t-1}|x_t)}{q(x_t|x_{t-1}, x_0)} \right]$$

$$= \mathbb{E}_{q(x_{1:T}|x_0)} \left[\log \frac{p(x_T) p_\theta(x_0|x_1)}{q(x_1|x_0)} + \log \prod_{t=2}^{T} \frac{p_\theta(x_{t-1}|x_t)}{\frac{q(x_{t-1}|x_t, x_0) q(x_t|x_0)}{q(x_{t-1}|x_0)}} \right]$$

Derivation of a Low-Variance ELBO II

$$= \mathbb{E}_{q(x_{1:T}|x_0)} \left[\log \frac{p(x_T) p_\theta(x_0|x_1)}{q(x_1|x_0)} + \log \prod_{t=2}^{T} \frac{p_\theta(x_{t-1}|x_t)}{\frac{q(x_{t-1}|x_t, x_0) \cancel{q(x_t|x_0)}}{\cancel{q(x_{t-1}|x_0)}}} \right]$$

$$= \mathbb{E}_{q(x_{1:T}|x_0)} \left[\log \frac{p(x_T) p_\theta(x_0|x_1)}{\cancel{q(x_1|x_0)}} + \log \frac{\cancel{q(x_1|x_0)}}{q(x_T|x_0)} + \log \prod_{t=2}^{T} \frac{p_\theta(x_{t-1}|x_t)}{q(x_{t-1}|x_t, x_0)} \right]$$

$$= \mathbb{E}_{q(x_{1:T}|x_0)} \left[\log \frac{p(x_T) p_\theta(x_0|x_1)}{q(x_T|x_0)} + \sum_{t=2}^{T} \log \frac{p_\theta(x_{t-1}|x_t)}{q(x_{t-1}|x_t, x_0)} \right]$$

$$= \mathbb{E}_{q(x_{1:T}|x_0)} [\log p_\theta(x_0|x_1)] + \mathbb{E}_{q(x_{1:T}|x_0)} \left[\log \frac{p(x_T)}{q(x_T|x_0)} \right] + \sum_{t=2}^{T} \mathbb{E}_{q(x_{1:T}|x_0)} \left[\log \frac{p_\theta(x_{t-1}|x_t)}{q(x_{t-1}|x_t, x_0)} \right]$$

$$= \mathbb{E}_{q(x_1|x_0)} [\log p_\theta(x_0|x_1)] + \mathbb{E}_{q(x_T|x_0)} \left[\log \frac{p(x_T)}{q(x_T|x_0)} \right] + \sum_{t=2}^{T} \mathbb{E}_{q(x_t, x_{t-1}|x_0)} \left[\log \frac{p_\theta(x_{t-1}|x_t)}{q(x_{t-1}|x_t, x_0)} \right]$$

$$= \underbrace{\mathbb{E}_{q(x_1|x_0)} [\log p_\theta(x_0|x_1)]}_{\text{reconstruction term}} - \underbrace{D_{\mathrm{KL}}(q(x_T|x_0) \| p(x_T))}_{\text{prior matching term}}$$

$$- \sum_{t=2}^{T} \underbrace{\mathbb{E}_{q(x_t|x_0)} [D_{\mathrm{KL}}(q(x_{t-1}|x_t, x_0) \| p_\theta(x_{t-1}|x_t))]}_{\text{denoising matching term}}$$

Interpretation of Each Term I

This formulation has an elegant interpretation, revealed when inspecting each individual term:

- Reconstruction Term:

$$\mathbb{E}_{q(\boldsymbol{x}_1|\boldsymbol{x}_0)}[\log p_{\boldsymbol{\theta}}(\boldsymbol{x}_0 \mid \boldsymbol{x}_1)]$$

This term is similar to its counterpart in a vanilla VAE and can be optimized using Monte Carlo estimates.

Interpretation of Each Term II

- Prior Matching Term:

$$D_{\mathrm{KL}}(q(\boldsymbol{x}_T \mid \boldsymbol{x}_0)\|p(\boldsymbol{x}_T))$$

Measures how closely the final noisified input distribution matches a standard Gaussian prior. No trainable parameters and zero under assumptions.

Interpretation of Each Term III

- Denoising Matching Term:

$$\mathbb{E}_{q(\boldsymbol{x}_t|\boldsymbol{x}_0)}[D_{\mathrm{KL}}(q(\boldsymbol{x}_{t-1} \mid \boldsymbol{x}_t, \boldsymbol{x}_0) \| p_{\boldsymbol{\theta}}(\boldsymbol{x}_{t-1} \mid \boldsymbol{x}_t))]$$

Minimized when the learned denoising transition $p_{\boldsymbol{\theta}}(\boldsymbol{x}_{t-1} \mid \boldsymbol{x}_t)$ matches the ground-truth denoising transition $q(\boldsymbol{x}_{t-1} \mid \boldsymbol{x}_t, \boldsymbol{x}_0)$.

Lower Variance Method to Optimize VDM

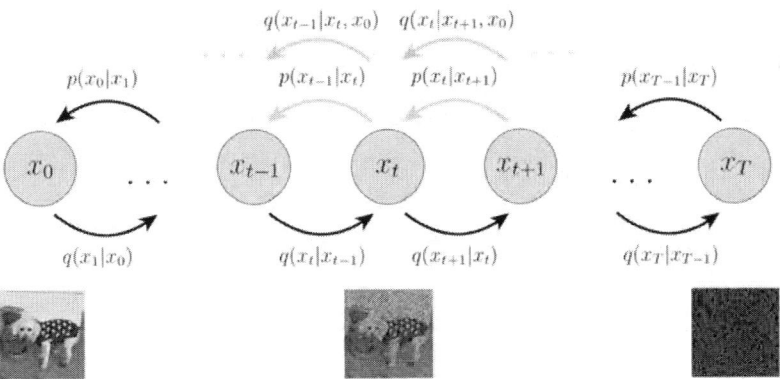

Figure: Alternate, lower-variance method to optimize a VDM

Optimizing Denoising Matching Term I

- Utilize Gaussian transition assumption for tractable optimization.
- Apply Bayes' Rule to express the conditional distribution:

$$q(\boldsymbol{x}_{t-1} \mid \boldsymbol{x}_t, \boldsymbol{x}_0) = \frac{q(\boldsymbol{x}_t \mid \boldsymbol{x}_{t-1}, \boldsymbol{x}_0)q(\boldsymbol{x}_{t-1} \mid \boldsymbol{x}_0)}{q(\boldsymbol{x}_t \mid \boldsymbol{x}_0)}$$

$q(\boldsymbol{x}_t \mid \boldsymbol{x}_{t-1}, \boldsymbol{x}_0)$

- Encoder Transitions Assumption:

$$q(\boldsymbol{x}_t \mid \boldsymbol{x}_0) = q(\boldsymbol{x}_t \mid \boldsymbol{x}_{t-1}) = \mathcal{N}(\boldsymbol{x}_t; \sqrt{\alpha_t}\boldsymbol{x}_{t-1}, (1-\alpha_t)\mathbf{I})$$

Variational Diffusion Models — Optimizing Denoising Matching Term
$q(x_t \mid x_0)$ I

- Reparameterization Trick for x_t:

$$x_t = \sqrt{\alpha_t}x_{t-1} + \sqrt{1-\alpha_t}\epsilon, \quad \epsilon \sim \mathcal{N}(0, I) \tag{13}$$

- Reparameterization Trick for x_{t-1}:

$$x_{t-1} = \sqrt{\alpha_{t-1}}x_{t-2} + \sqrt{1-\alpha_{t-1}}\epsilon, \quad \epsilon \sim \mathcal{N}(0, I) \tag{14}$$

- Recursively derive $q(x_t \mid x_0)$ using the reparameterization trick.

Variational Diffusion Models — Optimizing Denoising Matching Term
$q(x_t \mid x_0)$ II

Suppose we have access to $2T$ random noise variables $\{\epsilon_t^*, \epsilon_t\}_{t=0}^{T} \overset{\text{iid}}{\sim} \mathcal{N}(\epsilon; 0, I)$. Then, a sample $x_t \sim q(x_t \mid x_0)$ is

$$\begin{aligned}
x_t &= \sqrt{\alpha_t}x_{t-1} + \sqrt{1-\alpha_t}\epsilon_{t-1}^* \\
&= \sqrt{\alpha_t}\left(\sqrt{\alpha_{t-1}}x_{t-2} + \sqrt{1-\alpha_{t-1}}\epsilon_{t-2}^*\right) + \sqrt{1-\alpha_t}\epsilon_{t-1}^* \\
&= \sqrt{\alpha_t \alpha_{t-1}}x_{t-2} + \sqrt{\alpha_t - \alpha_t\alpha_{t-1}}\epsilon_{t-2}^* + \sqrt{1-\alpha_t}\epsilon_{t-1}^* \\
&= \sqrt{\alpha_t \alpha_{t-1}}x_{t-2} + \sqrt{\sqrt{\alpha_t - \alpha_t\alpha_{t-1}}^2 + \sqrt{1-\alpha_t}^2}\epsilon_{t-2} \\
&= \sqrt{\alpha_t \alpha_{t-1}}x_{t-2} + \sqrt{\alpha_t - \alpha_t\alpha_{t-1} + 1 - \alpha_t}\epsilon_{t-2} \\
&= \sqrt{\alpha_t \alpha_{t-1}}x_{t-2} + \sqrt{1 - \alpha_t\alpha_{t-1}}\epsilon_{t-2} \\
&= \ldots \\
&= \sqrt{\prod_{i=1}^{t}\alpha_i}\,x_0 + \sqrt{1 - \prod_{i=1}^{t}\alpha_i}\,\epsilon_0 \\
&= \sqrt{\bar{\alpha}_t}\,x_0 + \sqrt{1 - \bar{\alpha}_t}\,\epsilon_0 \\
&\sim \mathcal{N}\left(x_t; \sqrt{\bar{\alpha}_t}x_0, (1-\bar{\alpha}_t)I\right)
\end{aligned}$$

Variational Diffusion Models — Optimizing Denoising Matching Term
$q(x_t \mid x_0)$ III

- The sum of two independent Gaussian random variables is also a Gaussian. The new mean is the sum of the means, and the new variance is the sum of the variances.
- First Gaussian Sample: $\sqrt{1-\alpha_t}\epsilon^*_{t-1} \sim \mathcal{N}(0, (1-\alpha_t)\mathbf{I})$
- Second Gaussian Sample: $\sqrt{\alpha_t - \alpha_t\alpha_{t-1}}\epsilon^*_{t-2} \sim \mathcal{N}(0, (\alpha_t - \alpha_t\alpha_{t-1})\mathbf{I})$
- Sum of Both Samples:

$$\sim \mathcal{N}(0, (1-\alpha_t + \alpha_t - \alpha_t\alpha_{t-1})\mathbf{I}) = \mathcal{N}(0, (1-\alpha_t\alpha_{t-1})\mathbf{I})$$

- Reparameterization Trick for Sum:

$$\sqrt{1-\alpha_t\alpha_{t-1}}\,\epsilon_{t-2}$$

Variational Diffusion Models — Optimizing Denoising Matching Term
$q(x_{t-1} \mid x_t, x_0)$ I

$$
\begin{aligned}
&q(x_{t-1} \mid x_t, x_0) \\
&= \frac{q(x_t \mid x_{t-1}, x_0)\, q(x_{t-1} \mid x_0)}{q(x_t \mid x_0)} \\
&= \frac{\mathcal{N}(x_t;\sqrt{\alpha_t}x_{t-1},(1-\alpha_t)\mathbf{I})\,\mathcal{N}(x_{t-1};\sqrt{\bar{\alpha}_{t-1}}x_0,(1-\bar{\alpha}_{t-1})\mathbf{I})}{\mathcal{N}(x_t;\sqrt{\bar{\alpha}_t}x_0,(1-\bar{\alpha}_t)\mathbf{I})} \\
&\propto \exp\left\{-\left[\frac{(x_t-\sqrt{\alpha_t}x_{t-1})^2}{2(1-\alpha_t)} + \frac{(x_{t-1}-\sqrt{\bar{\alpha}_{t-1}}x_0)^2}{2(1-\bar{\alpha}_{t-1})} - \frac{(x_t-\sqrt{\bar{\alpha}_t}x_0)^2}{2(1-\bar{\alpha}_t)}\right]\right\} \\
&= \exp\left\{-\frac{1}{2}\left[\frac{(x_t-\sqrt{\alpha_t}x_{t-1})^2}{1-\alpha_t} + \frac{(x_{t-1}-\sqrt{\bar{\alpha}_{t-1}}x_0)^2}{1-\bar{\alpha}_{t-1}} - \frac{(x_t-\sqrt{\bar{\alpha}_t}x_0)^2}{1-\bar{\alpha}_t}\right]\right\} \\
&= \exp\left\{-\frac{1}{2}\left[\frac{(-2\sqrt{\alpha_t}x_t x_{t-1} + \alpha_t x_{t-1}^2)}{1-\alpha_t} + \frac{(x_{t-1}^2 - 2\sqrt{\bar{\alpha}_{t-1}}x_{t-1}x_0)}{1-\bar{\alpha}_{t-1}} + C(x_t, x_0)\right]\right\} \\
&\propto \exp\left\{-\frac{1}{2}\left[-\frac{2\sqrt{\alpha_t}x_t x_{t-1}}{1-\alpha_t} + \frac{\alpha_t x_{t-1}^2}{1-\alpha_t} + \frac{x_{t-1}^2}{1-\bar{\alpha}_{t-1}} - \frac{2\sqrt{\bar{\alpha}_{t-1}}x_{t-1}x_0}{1-\bar{\alpha}_{t-1}}\right]\right\} \\
&= \exp\left\{-\frac{1}{2}\left[\left(\frac{\alpha_t}{1-\alpha_t} + \frac{1}{1-\bar{\alpha}_{t-1}}\right)x_{t-1}^2 - 2\left(\frac{\sqrt{\alpha_t}x_t}{1-\alpha_t} + \frac{\sqrt{\bar{\alpha}_{t-1}}x_0}{1-\bar{\alpha}_{t-1}}\right)x_{t-1}\right]\right\}
\end{aligned}
$$

Variational Diffusion Models — Optimizing Denoising Matching Term

$q\left(\boldsymbol{x}_{t-1} \mid \boldsymbol{x}_t, \boldsymbol{x}_0\right) \parallel$

$$= \exp\left\{-\frac{1}{2}\left[\frac{\alpha_t\left(1-\bar{\alpha}_{t-1}\right)+1-\alpha_t}{\left(1-\alpha_t\right)\left(1-\bar{\alpha}_{t-1}\right)}\boldsymbol{x}_{t-1}^2 - 2\left(\frac{\sqrt{\alpha_t}\boldsymbol{x}_t}{1-\alpha_t}+\frac{\sqrt{\bar{\alpha}_{t-1}}\boldsymbol{x}_0}{1-\bar{\alpha}_{t-1}}\right)\boldsymbol{x}_{t-1}\right]\right\}$$

$$= \exp\left\{-\frac{1}{2}\left[\frac{\alpha_t-\bar{\alpha}_t+1-\alpha_t}{\left(1-\alpha_t\right)\left(1-\bar{\alpha}_{t-1}\right)}\boldsymbol{x}_{t-1}^2 - 2\left(\frac{\sqrt{\alpha_t}\boldsymbol{x}_t}{1-\alpha_t}+\frac{\sqrt{\bar{\alpha}_{t-1}}\boldsymbol{x}_0}{1-\bar{\alpha}_{t-1}}\right)\boldsymbol{x}_{t-1}\right]\right\}$$

$$= \exp\left\{-\frac{1}{2}\left[\frac{1-\bar{\alpha}_t}{\left(1-\alpha_t\right)\left(1-\bar{\alpha}_{t-1}\right)}\boldsymbol{x}_{t-1}^2 - 2\left(\frac{\sqrt{\alpha_t}\boldsymbol{x}_t}{1-\alpha_t}+\frac{\sqrt{\bar{\alpha}_{t-1}}\boldsymbol{x}_0}{1-\bar{\alpha}_{t-1}}\right)\boldsymbol{x}_{t-1}\right]\right\}$$

$$= \exp\left\{-\frac{1}{2}\left(\frac{1-\bar{\alpha}_t}{\left(1-\alpha_t\right)\left(1-\bar{\alpha}_{t-1}\right)}\right)\left[\boldsymbol{x}_{t-1}^2 - 2\frac{\left(\frac{\sqrt{\alpha_t}\boldsymbol{x}_t}{1-\alpha_t}+\frac{\sqrt{\bar{\alpha}_{t-1}}\boldsymbol{x}_0}{1-\bar{\alpha}_{t-1}}\right)}{\frac{1-\bar{\alpha}_t}{\left(1-\alpha_t\right)\left(1-\bar{\alpha}_{t-1}\right)}}\boldsymbol{x}_{t-1}\right]\right\}$$

$$= \exp\left\{-\frac{1}{2}\left(\frac{1-\bar{\alpha}_t}{\left(1-\alpha_t\right)\left(1-\bar{\alpha}_{t-1}\right)}\right)\left[\boldsymbol{x}_{t-1}^2 - 2\frac{\left(\frac{\sqrt{\alpha_t}\boldsymbol{x}_t}{1-\alpha_t}+\frac{\sqrt{\bar{\alpha}_{t-1}}\boldsymbol{x}_0}{1-\bar{\alpha}_{t-1}}\right)\left(1-\alpha_t\right)\left(1-\bar{\alpha}_{t-1}\right)}{1-\bar{\alpha}_t}\boldsymbol{x}_{t-1}\right]\right\}$$

$$= \exp\left\{-\frac{1}{2}\left(\frac{1}{\frac{\left(1-\alpha_t\right)\left(1-\bar{\alpha}_{t-1}\right)}{1-\bar{\alpha}_t}}\right)\left[\boldsymbol{x}_{t-1}^2 - 2\frac{\sqrt{\alpha_t}\left(1-\bar{\alpha}_{t-1}\right)\boldsymbol{x}_t + \sqrt{\bar{\alpha}_{t-1}}\left(1-\alpha_t\right)\boldsymbol{x}_0}{1-\bar{\alpha}_t}\boldsymbol{x}_{t-1}\right]\right\}$$

$$\propto \mathcal{N}(\boldsymbol{x}_{t-1}; \underbrace{\frac{\sqrt{\alpha_t}\left(1-\bar{\alpha}_{t-1}\right)\boldsymbol{x}_t + \sqrt{\bar{\alpha}_{t-1}}\left(1-\alpha_t\right)\boldsymbol{x}_0}{1-\bar{\alpha}_t}}_{\boldsymbol{\mu}_q(\boldsymbol{x}_t,\boldsymbol{x}_0)}, \underbrace{\frac{\left(1-\alpha_t\right)\left(1-\bar{\alpha}_{t-1}\right)}{1-\bar{\alpha}_t}\boldsymbol{I}}_{\boldsymbol{\Sigma}_q(t)})$$

Variational Diffusion Models — Optimizing Denoising Matching Term

$p_{\boldsymbol{\theta}}(\boldsymbol{x}_{t-1} \mid \boldsymbol{x}_t)$

- Distribution of $\boldsymbol{x}_{t-1}$:
  - $\boldsymbol{x}_{t-1} \sim q(\boldsymbol{x}_{t-1} \mid \boldsymbol{x}_t, \boldsymbol{x}_0)$ is Gaussian with mean $\boldsymbol{\mu}_q(\boldsymbol{x}_t, \boldsymbol{x}_0)$ and variance $\boldsymbol{\Sigma}_q(t)$.
- Variance $\boldsymbol{\Sigma}_q(t)$:
$$\sigma_q^2(t) = \frac{(1-\alpha_t)(1-\bar{\alpha}_{t-1})}{1-\bar{\alpha}_t}$$
- α-Coefficients:
  - Known and fixed at each time step.
  - Either set as hyperparameters or inferred by a network.
- Approximate Denoising Transition Step $p_{\boldsymbol{\theta}}(\boldsymbol{x}_{t-1} \mid \boldsymbol{x}_t)$:
  - Modeled as Gaussian.
  - Variance is the same as $q(\boldsymbol{x}_{t-1} \mid \boldsymbol{x}_t, \boldsymbol{x}_0)$, i.e., $\boldsymbol{\Sigma}_q(t) = \sigma_q^2(t)\boldsymbol{I}$.
  - Mean $\boldsymbol{\mu}_{\boldsymbol{\theta}}(\boldsymbol{x}_t, t)$ is parameterized as a function of $\boldsymbol{x}_t$ since it does not condition on $\boldsymbol{x}_0$.

KL Divergence between Two Gaussian Distributions

$$D_{\mathrm{KL}}\left(\mathcal{N}\left(\boldsymbol{x};\boldsymbol{\mu}_x,\boldsymbol{\Sigma}_x\right) \parallel \mathcal{N}\left(\boldsymbol{y};\boldsymbol{\mu}_y,\boldsymbol{\Sigma}_y\right)\right)$$
$$= \frac{1}{2}\left[\log\frac{|\boldsymbol{\Sigma}_y|}{|\boldsymbol{\Sigma}_x|} - d + \mathrm{tr}\left(\boldsymbol{\Sigma}_y^{-1}\boldsymbol{\Sigma}_x\right) + \left(\boldsymbol{\mu}_y - \boldsymbol{\mu}_x\right)^T \boldsymbol{\Sigma}_y^{-1}\left(\boldsymbol{\mu}_y - \boldsymbol{\mu}_x\right)\right]$$

Denoising Matching Term

Optimizing the KL Divergence term reduces to minimizing the difference between the means of the two distributions:

$$\arg\min_\theta D_{\mathrm{KL}}\left(q\left(\boldsymbol{x}_{t-1} \mid \boldsymbol{x}_t, \boldsymbol{x}_0\right) \parallel p_\theta\left(\boldsymbol{x}_{t-1} \mid \boldsymbol{x}_t\right)\right)$$
$$=\arg\min_\theta D_{\mathrm{KL}}\left(\mathcal{N}\left(\boldsymbol{x}_{t-1};\boldsymbol{\mu}_q,\boldsymbol{\Sigma}_q(t)\right) \parallel \mathcal{N}\left(\boldsymbol{x}_{t-1};\boldsymbol{\mu}_\theta,\boldsymbol{\Sigma}_q(t)\right)\right)$$
$$=\arg\min_\theta \frac{1}{2}\left[\log\frac{|\boldsymbol{\Sigma}_q(t)|}{|\boldsymbol{\Sigma}_q(t)|} - d + \mathrm{tr}\left(\boldsymbol{\Sigma}_q(t)^{-1}\boldsymbol{\Sigma}_q(t)\right) + \left(\boldsymbol{\mu}_\theta - \boldsymbol{\mu}_q\right)^T \boldsymbol{\Sigma}_q(t)^{-1}\left(\boldsymbol{\mu}_\theta - \boldsymbol{\mu}_q\right)\right]$$
$$=\arg\min_\theta \frac{1}{2}\left[\log 1 - d + d + \left(\boldsymbol{\mu}_\theta - \boldsymbol{\mu}_q\right)^T \boldsymbol{\Sigma}_q(t)^{-1}\left(\boldsymbol{\mu}_\theta - \boldsymbol{\mu}_q\right)\right]$$
$$=\arg\min_\theta \frac{1}{2}\left[\left(\boldsymbol{\mu}_\theta - \boldsymbol{\mu}_q\right)^T \boldsymbol{\Sigma}_q(t)^{-1}\left(\boldsymbol{\mu}_\theta - \boldsymbol{\mu}_q\right)\right]$$
$$=\arg\min_\theta \frac{1}{2}\left[\left(\boldsymbol{\mu}_\theta - \boldsymbol{\mu}_q\right)^T \left(\sigma_q^2(t)\mathbf{I}\right)^{-1}\left(\boldsymbol{\mu}_\theta - \boldsymbol{\mu}_q\right)\right]$$
$$=\arg\min_\theta \frac{1}{2\sigma_q^2(t)}\left[\|\boldsymbol{\mu}_\theta - \boldsymbol{\mu}_q\|_2^2\right]$$

where we have written $\boldsymbol{\mu}_q$ as shorthand for $\boldsymbol{\mu}_q\left(\boldsymbol{x}_t, \boldsymbol{x}_0\right)$, and $\boldsymbol{\mu}_\theta$ as shorthand for $\boldsymbol{\mu}_\theta\left(\boldsymbol{x}_t, t\right)$ for brevity.

μ_θ

Optimize a $\mu_\theta(x_t, t)$ that matches $\mu_q(x_t, x_0)$, which takses the form:

$$\mu_q(x_t, x_0) = \frac{\sqrt{\alpha_t}(1 - \bar{\alpha}_{t-1})x_t + \sqrt{\bar{\alpha}_{t-1}}(1 - \alpha_t)x_0}{1 - \bar{\alpha}_t}$$

As $\mu_\theta(x_t, t)$ also conditions on x_t, we can match $\mu_q(x_t, x_0)$ closely by setting it to the following form:

$$\mu_\theta(x_t, t) = \frac{\sqrt{\alpha_t}(1 - \bar{\alpha}_{t-1})x_t + \sqrt{\bar{\alpha}_{t-1}}(1 - \alpha_t)\hat{x}_\theta(x_t, t)}{1 - \bar{\alpha}_t}$$

where $\hat{x}_\theta(x_t, t)$ is parameterized by a neural network that seeks to predict x_0 from noisy image x_t and time index t.

Simplified KL Term

$$\begin{aligned}
&\arg\min_\theta D_{\mathrm{KL}}(q(x_{t-1} \mid x_t, x_0) \parallel p_\theta(x_{t-1} \mid x_t)) \\
=&\arg\min_\theta D_{\mathrm{KL}}(\mathcal{N}(x_{t-1}; \mu_q, \Sigma_q(t)) \parallel \mathcal{N}(x_{t-1}; \mu_\theta, \Sigma_q(t))) \\
=&\arg\min_\theta \frac{1}{2\sigma_q^2(t)}\left[\left\|\frac{\sqrt{\alpha_t}(1-\bar{\alpha}_{t-1})x_t + \sqrt{\bar{\alpha}_{t-1}}(1-\alpha_t)\hat{x}_\theta(x_t,t)}{1-\bar{\alpha}_t}\right.\right.\\
&\qquad\qquad\left.\left. - \frac{\sqrt{\alpha_t}(1-\bar{\alpha}_{t-1})x_t + \sqrt{\bar{\alpha}_{t-1}}(1-\alpha_t)x_0}{1-\bar{\alpha}_t}\right\|_2^2\right] \\
=&\arg\min_\theta \frac{1}{2\sigma_q^2(t)}\left[\left\|\frac{\sqrt{\bar{\alpha}_{t-1}}(1-\alpha_t)\hat{x}_\theta(x_t,t)}{1-\bar{\alpha}_t} - \frac{\sqrt{\bar{\alpha}_{t-1}}(1-\alpha_t)x_0}{1-\bar{\alpha}_t}\right\|_2^2\right] \\
=&\arg\min_\theta \frac{1}{2\sigma_q^2(t)}\left[\left\|\frac{\sqrt{\bar{\alpha}_{t-1}}(1-\alpha_t)}{1-\bar{\alpha}_t}(\hat{x}_\theta(x_t,t) - x_0)\right\|_2^2\right] \\
=&\arg\min_\theta \frac{1}{2\sigma_q^2(t)}\frac{\bar{\alpha}_{t-1}(1-\alpha_t)^2}{(1-\bar{\alpha}_t)^2}\left[\|\hat{x}_\theta(x_t,t) - x_0\|_2^2\right]
\end{aligned}$$

Interpreting VDM Optimization

Variational Diffusion Models — Optimizing Denoising Matching Term

- VDM Optimization Objective:
  - Learning a neural network to predict the original ground truth image from its noisified version.
- Derived ELBO Objective:

$$\arg\min_{\boldsymbol{\theta}} \mathbb{E}_{t \sim U\{2,T\}} \left[\mathbb{E}_{q(\boldsymbol{x}_t \mid \boldsymbol{x}_0)} \left[D_{\mathrm{KL}}\left(q\left(\boldsymbol{x}_{t-1} \mid \boldsymbol{x}_t, \boldsymbol{x}_0\right) \;\|\; p_{\boldsymbol{\theta}}\left(\boldsymbol{x}_{t-1} \mid \boldsymbol{x}_t\right) \right) \right] \right]$$

- Optimization Method:
  - Stochastic sampling over timesteps is used for optimization.

강 의 노 트

🚗 1.

🚗 2.

🚗 3.

Chapter 10.

Generative Adversarial Networks

Introduction

- Focus on *implicit generative models* without an explicit likelihood function.
- Includes family of Generative Adversarial Networks (GANs).

Two Types of Probabilistic Models

- **Prescribed (Explicit) Models:** Provide explicit parametric distribution of observed x.

$$\log q_\theta(x)$$

- **Implicit Models:** Define a stochastic procedure to generate data directly.

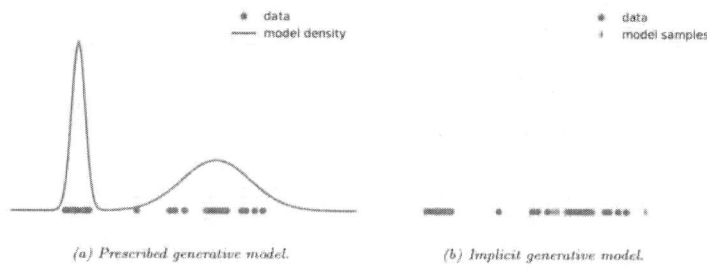

(a) Prescribed generative model.　(b) Implicit generative model.

Figure: Difference between prescribed and implicit models.

Learning by Comparison

In VAEs, we rely on the principle of maximum likelihood for learning.

- Maximize likelihood $\Rightarrow$ Minimize KL divergence
- Model: q_θ
- True data distribution: p^*

Implicit models only provide samples. Therefore, need new learning principles.

Overview of Approaches

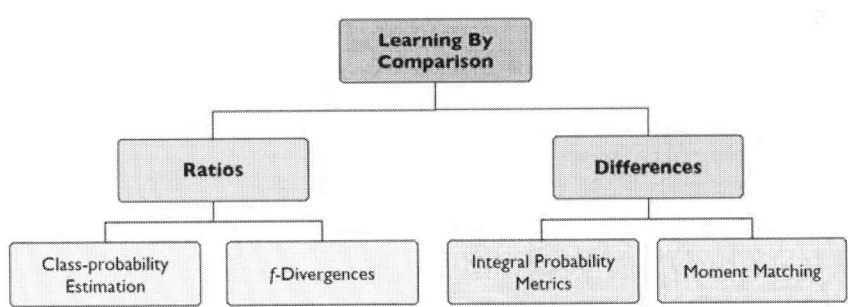

Figure: Overview of approaches for learning in implicit generative models

Learning in Implicit Models

- Determine if distributions are close from two sets of samples.
- Quantify distance: 'two sample' or likelihood-free approach.
- Methods: Distributional divergences, distances through binary classification, method of moments, etc.

Guiding Principles

Looking for objectives $\mathcal{D}(p^*, q)$ with:

1. Data distribution guarantee: $\operatorname{argmin}_q \mathcal{D}(p^*, q) = p^*$.
2. Evaluated only using samples.
3. Computationally cheap.

Many distributional distances and divergences (e.g., KL divergence) satisfy the first requirement:

$$\mathcal{D}(p^*, q) \geq 0; \quad \mathcal{D}(p^*, q) = 0 \iff p^* = q.$$

However, they fail to satisfy the other two requirements: they cannot be evaluated only using samples.

Approach to evaluate only using samples

- Approximate via optimization; use a discriminator or critic D.
- Objective depends on samples: $\mathcal{F}(D_\phi, p^*, q_\theta)$.

$$\mathcal{D}(p^*, q) = \underset{D}{\operatorname{argmax}} \, \mathcal{F}(D, p^*, q)$$

where $\mathcal{F}$ is a functional that depends on p^* and q only through samples.

Optimize over Parameters Instead of Functions

- The model q_θ and the critic D are parametric with parameters $\boldsymbol{\theta}$ and ϕ respectively.
- Therefore, optimization wrt function D can be reposed as optimization wrt parameters. For the critic, this results in the optimization problem

$$\operatorname{argmax}_\phi \mathcal{F}(D_\phi, p^*, q_\theta)$$

- For the model parameters $\boldsymbol{\theta}$, the exact objective $\mathcal{D}(p^*, q_\theta)$ is replaced with the tractable approximation provided through the use of D_ϕ:

Approach for Estimation Using Samples

- Depend on the two distributions only in expectation.

$$\mathcal{F}(D_\phi, p^*, q_\theta) = \mathbb{E}_{p^*(x)} f(x, \phi) + \mathbb{E}_{q_\theta(x)} g(x, \phi)$$

where f and g are real valued functions whose choice will define $\mathcal{F}$.
- Rewriting for implicit models:

$$\mathcal{F}(D_\phi, p^*, q_\theta) = \mathbb{E}_{p^*(x)} f(x, \phi) + \mathbb{E}_{q(z)} g(G_\theta(z), \phi) \quad (15)$$

where $x = G_\theta(z), z \sim q(z)$

Guiding Principles for Instantiation

- Next: Finding the functions f and g, and thus the objective $\mathcal{F}$.
- Methods for instantiation:
    - Class probability estimation
    - Bounds on f-divergences
    - Integral probability metrics
    - Moment matching

Density Ratio Estimation Using Binary Classifiers

- Compare two distributions p^* and q_θ by computing density ratio $r(\boldsymbol{x}) = \frac{p^*(\boldsymbol{x})}{q_\theta(\boldsymbol{x})}$.
- Converts density estimation into a binary classification problem:
  - Points from p^* have label $y = 1$ ($p(\boldsymbol{x} \mid y = 1)$)
  - Points from q_θ have label $y = 0$ ($p(\boldsymbol{x} \mid y = 0)$)
  - Let $p(y = 1) = \pi$ be the class prior
  - By Bayes' rule, the density ratio $r(\boldsymbol{x}) = P(\boldsymbol{x})/q_\theta(\boldsymbol{x})$ is given by

$$\frac{p^*(\boldsymbol{x})}{q_\theta(\boldsymbol{x})} = \frac{p(\boldsymbol{x} \mid y = 1)}{p(\boldsymbol{x} \mid y = 0)} = \frac{p(y = 1 \mid \boldsymbol{x})p(\boldsymbol{x})}{p(y = 1)} \bigg/ \frac{p(y = 0 \mid \boldsymbol{x})p(\boldsymbol{x})}{p(y = 0)}$$

$$= \frac{p(y = 1 \mid \boldsymbol{x})}{p(y = 0 \mid \boldsymbol{x})} \frac{1 - \pi}{\pi}$$

- By letting $\pi = \frac{1}{2}$, $D(\boldsymbol{x}) = p(y = 1 \mid \boldsymbol{x})$,

$$\frac{p^*(\boldsymbol{x})}{q_\theta(\boldsymbol{x})} = \frac{D(\boldsymbol{x})}{1 - D(\boldsymbol{x})}$$

Discriminator or Critic

- $D(\boldsymbol{x})$: Trained to distinguish samples from p^* vs q_θ.
- For parametric classification, learn $D_\phi(\boldsymbol{x}) \in [0, 1]$ with parameters ϕ.
- For the Bernoulli log-loss (or binary cross entropy loss), we obtain the objective:

$$V(q_\theta, p^*) = \arg\max_\phi \mathbb{E}_{p(\boldsymbol{x}|y)p(y)} [y \log D_\phi(\boldsymbol{x}) + (1 - y) \log(1 - D_\phi(\boldsymbol{x}))]$$

$$= \arg\max_\phi \mathbb{E}_{p(\boldsymbol{x}|y=1)p(y=1)} \log D_\phi(\boldsymbol{x}) + \mathbb{E}_{p(\boldsymbol{x}|y=0)p(y=0)} \log(1 - D_\phi(\boldsymbol{x}))$$

$$= \arg\max_\phi \frac{1}{2}\mathbb{E}_{p^*(\boldsymbol{x})} \log D_\phi(\boldsymbol{x}) + \frac{1}{2}\mathbb{E}_{q_\theta(\boldsymbol{x})} \log(1 - D_\phi(\boldsymbol{x})) \quad (16)$$

Optimal discriminator

- The optimal discriminator D is $\frac{p^*(x)}{p^*(x)+q_\theta(x)}$, since:

$$\frac{p^*(x)}{q_\theta(x)} = \frac{D^*(x)}{1-D^*(x)} \implies D^*(x) = \frac{p^*(x)}{p^*(x)+q_\theta(x)}$$

Jensen-Shannon Divergence

Substituting the optimal discriminator into the scoring rule (16):

$$\begin{aligned} V^*(q_\theta, p^*) &= \frac{1}{2}\mathbb{E}_{p^*(x)}\left[\log \frac{p^*(x)}{p^*(x)+q_\theta(x)}\right] + \frac{1}{2}\mathbb{E}_{q_\theta(x)}\left[\log\left(1-\frac{p^*(x)}{p^*(x)+q_\theta(x)}\right)\right] \\ &= \frac{1}{2}\mathbb{E}_{p^*(x)}\left[\log \frac{p^*(x)}{\frac{p^*(x)+q_\theta(x)}{2}}\right] + \frac{1}{2}\mathbb{E}_{q_\theta(x)}\left[\log\left(\frac{q_\theta(x)}{\frac{p^*(x)+q_\theta(x)}{2}}\right)\right] - \log 2 \\ &= \frac{1}{2}D_{\mathrm{KL}}\left(p^* \| \frac{p^*+q_\theta}{2}\right) + \frac{1}{2}D_{\mathrm{KL}}\left(q_\theta \| \frac{p^*+q_\theta}{2}\right) - \log 2 \\ &= JSD(p^*, q_\theta) - \log 2 \end{aligned}$$

where JSD denotes the Jensen-Shannon divergence:

$$JSD(p^*, q_\theta) = \frac{1}{2}D_{\mathrm{KL}}\left(p^* \| \frac{p^*+q_\theta}{2}\right) + \frac{1}{2}D_{\mathrm{KL}}\left(q_\theta \| \frac{p^*+q_\theta}{2}\right)$$

Training θ

Train the parameters θ of generative model to minimize the divergence:

$$\min_{\theta} JSD\left(p^*, q_\theta\right) = \min_{\theta} V^*\left(q_\theta, p^*\right) + \log 2$$

$$= \min_{\theta} \frac{1}{2}\mathbb{E}_{p^*(x)} \log D^*(x) + \frac{1}{2}\mathbb{E}_{q_\theta(x)} \log\left(1 - D^*(x)\right) + \log 2$$

Final Form of Objective

Since we do not have access to the optimal classifier D^* but only to the neural approximation D_ϕ obtained using the optimization in (16), this results in a min-max optimization problem:

$$\min_{\theta} \max_{\phi} \frac{1}{2}\mathbb{E}_{p^*(x)}\left[\log D_\phi(x)\right] + \frac{1}{2}\mathbb{E}_{q_\theta(x)}\left[\log\left(1 - D_\phi(x)\right)\right]$$

Objective in terms of the latent variables z of the implicit generative model:

$$\min_{\theta} \max_{\phi} \frac{1}{2}\mathbb{E}_{p^*(x)}\left[\log D_\phi(x)\right] + \frac{1}{2}\mathbb{E}_{q(z)}\left[\log\left(1 - D_\phi\left(G_\theta(z)\right)\right)\right]$$

f-divergences

- General class of divergences: f-divergences.

$$\mathcal{D}_f[p^*(\boldsymbol{x})\|q_\theta(\boldsymbol{x})] = \int q_\theta(\boldsymbol{x}) f\left(\frac{p^*(\boldsymbol{x})}{q_\theta(\boldsymbol{x})}\right) d\boldsymbol{x} \qquad (17)$$

- f is a convex function such that $f(1) = 0$.
- Different choices of f: KL, reverse KL, Jensen-Shannon divergence.

Table of Standard Divergences

Divergence	f	$f^\dagger$	Optimal Critic
KL	$u \log u$	e^{u-1}	$1 + \log r(\boldsymbol{x})$
Reverse KL	$-\log u$	$-1 - \log(-u)$	$-1/r(\boldsymbol{x})$
JSD	$u \log u - (u+1)\log\frac{u+1}{2}$	$-\log(2 - e^u)$	$\frac{2}{1+1/r(x)}$
Pearson χ^2	$(u-1)^2$	$\frac{1}{4}u^2 + u$	$(\sqrt{r(\boldsymbol{x})} - 1)\sqrt{1/r(\boldsymbol{x})}$

Table: Standard divergences as f divergences for various choices of f. The optimal critic is written as a function of the density ratio $r(\boldsymbol{x}) = \frac{p^*(x)}{q_\theta(x)}$.

Conjugate duality[1]

- To use f-divergences as a two-sample test objective for likelihood-free learning, we need to be able to estimate it only via samples
- Fenchel conjugate: For any function $f(\cdot)$, its convex conjugate is defined as

$$f^*(t) = \sup_{x \in \text{dom}_f} (xt - f(x))$$

- f^* is always convex and lower semi-continuous.
- $f^{**} \leq f$.
- Duality: $f^{**} = f$ when $f(\cdot)$ is convex, lower semicontinous. Or,

$$f(x) = f^{**}(x) = \sup_{t \in \text{dom}_{f^*}} (tx - f^*(t))$$

---
[1] Slide from Stanford CS236

Conjugate duality (Cont'd)

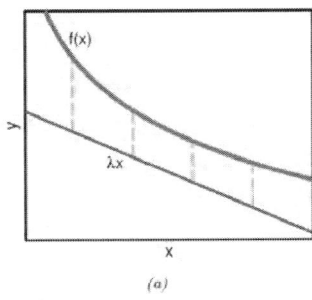 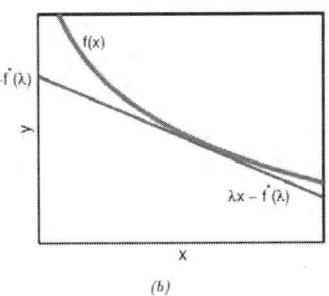

(a) (b)

Figure 6.4: Illustration of a conjugate functon. Red line is original function $f(x)$, and the blue line is a linear lower bound λx. To make the bound tight, we find the x where $\nabla f(x)$ is parallel to λ, and slide the line up to touch there; the amount we slide up is given by $f^*(\lambda)$. Adapted from Figure 10.11 of [BishopBook].

Variational Formulation of f-Divergences

Variational formulation provides lower bounds on f-divergences:

$$\begin{aligned}
\mathcal{D}_f\left[p^*(\boldsymbol{x})\|q_\theta(\boldsymbol{x})\right] &= \int q_\theta(\boldsymbol{x}) f\left(\frac{p^*(\boldsymbol{x})}{q_\theta(\boldsymbol{x})}\right) d\boldsymbol{x} \\
&= \int q_\theta(\boldsymbol{x}) \sup_{t:\mathcal{X}\to\mathbb{R}} \left[t(\boldsymbol{x})\frac{p^*(\boldsymbol{x})}{q_\theta(\boldsymbol{x})} - f^\dagger(t(\boldsymbol{x}))\right] d\boldsymbol{x} \\
&= \int \sup_{t:\mathcal{X}\to\mathbb{R}} p^*(\boldsymbol{x})t(\boldsymbol{x}) - q_\theta(\boldsymbol{x}) f^\dagger(t(\boldsymbol{x})) d\boldsymbol{x} \\
&\geq \sup_{t\in\mathcal{T}} \mathbb{E}_{p^*(\boldsymbol{x})}[t(\boldsymbol{x})] - \mathbb{E}_{q_\theta(\boldsymbol{x})}\left[f^\dagger(t(\boldsymbol{x}))\right].
\end{aligned}$$

where $f^\dagger$ is the convex conjugate of f.

Bounded Objective

Replacing t with D_ϕ:

$$\begin{aligned}
\min_{\boldsymbol{\theta}} \mathcal{D}_f(p^*, q_\theta) &\geq \min_{\boldsymbol{\theta}} \sup_{t\in\mathcal{T}} \mathbb{E}_{p^*(\boldsymbol{x})}[t(\boldsymbol{x})] - \mathbb{E}_{q_\theta(\boldsymbol{x})}\left[f^\dagger(t(\boldsymbol{x}))\right] \\
&\geq \min_{\boldsymbol{\theta}} \max_{\phi} \mathbb{E}_{p^*(\boldsymbol{x})}[D_\phi(\boldsymbol{x})] - \mathbb{E}_{q_\theta(\boldsymbol{x})}\left[f^\dagger(D_\phi(\boldsymbol{x}))\right] \\
&= \min_{\boldsymbol{\theta}} \max_{\phi} \mathbb{E}_{p^*(\boldsymbol{x})}[D_\phi(\boldsymbol{x})] - \mathbb{E}_{q(\boldsymbol{z})}\left[f^\dagger(D_\phi(G_{\boldsymbol{\theta}}(\boldsymbol{z})))\right]
\end{aligned}$$

- Training both the generator and critic parameters.
- Leading to f-GANs.

Integral Probability Metrics

- Definition:

$$I_{\mathcal{F}}(p^*(\boldsymbol{x}), q_\theta(\boldsymbol{x})) = \sup_{f \in \mathcal{F}} \left| \mathbb{E}_{p^*(\boldsymbol{x})} f(\boldsymbol{x}) - \mathbb{E}_{q_\theta(\boldsymbol{x})} f(\boldsymbol{x}) \right|$$

where the function f is a test or witness function that will take the role of the discriminator or critic.

- Valid function family $\mathcal{F}$ should satisfy

$$\mathcal{D}(p^*, q) \geq 0$$
$$\mathcal{D}(p^*, q) = 0 \iff p^* = q$$

Choosing the Function Family $\mathcal{F}$

Wasserstein Distance:

$$\begin{aligned} W_1(p^*(\boldsymbol{x}), q_\theta(\boldsymbol{x})) &= \sup_{f: \|f\|_{\text{Lip}} \leq 1} \mathbb{E}_{p^*(\boldsymbol{x})} f(\boldsymbol{x}) - \mathbb{E}_{q_\theta(\boldsymbol{x})} f(\boldsymbol{x}) \\ &\geq \max_{\phi: \|D_\phi\|_{\text{Lip}} \leq 1} \mathbb{E}_{p^*(\boldsymbol{x})} D_\phi(\boldsymbol{x}) - \mathbb{E}_{q_\theta(\boldsymbol{x})} D_\phi(\boldsymbol{x}) \end{aligned}$$

where the critic D_ϕ has to be regularized to be 1-Lipschitz. 1-Lipschitz can be enforced via

- Gradient penalties
- Spectral normalization

(Appendix) Wasserstein GAN: beyond f-divergences[2]

The f-divergence is defined as

$$D_f(p,q) = E_{\mathbf{x} \sim q}\left[f\left(\frac{p(\mathbf{x})}{q(\mathbf{x})} \right) \right]$$

- The support of q has to cover the support of p. Otherwise discontinuity arises in f-divergences.
- Let $p(\mathbf{x}) = \begin{cases} 1, & \mathbf{x} = 0 \\ 0, & \mathbf{x} \neq 0 \end{cases}$, and $q_\theta(\mathbf{x}) = \begin{cases} 1, & \mathbf{x} = \theta \\ 0, & \mathbf{x} \neq \theta \end{cases}$.
- $D_{KL}(p, q_\theta) = \begin{cases} 0, & \theta = 0 \\ \infty, & \theta \neq 0 \end{cases}$.
- $D_{JS}(p, q_\theta) = \begin{cases} 0, & \theta = 0 \\ \log 2, & \theta \neq 0 \end{cases}$.
- We need a "smoother" distance $D(p, q)$ that is defined when p and q have disjoint supports.

[2]Slide from Stanford CS 236

(Appendix) Wasserstein (Earth-Mover) distance[3]

- Wasserstein distance:

$$D_w(p, q) = \inf_{\gamma \in \Pi(p,q)} E_{(\mathbf{x},\mathbf{y}) \sim \gamma}[\|\mathbf{x} - \mathbf{y}\|_1]$$

where $\Pi(p, q)$ contains all joint distributions of $(\mathbf{x}, \mathbf{y})$ where the marginal of $\mathbf{x}$ is $p(\mathbf{x})$, and the marginal of $\mathbf{y}$ is $q(\mathbf{y})$.
- $\gamma(\mathbf{y} \mid \mathbf{x})$: a probabilistic earth moving plan that warps $p(\mathbf{x})$ to $q(\mathbf{y})$.
- Let $p(\mathbf{x}) = \begin{cases} 1, & \mathbf{x} = 0 \\ 0, & \mathbf{x} \neq 0 \end{cases}$, and $q_\theta(\mathbf{x}) = \begin{cases} 1, & \mathbf{x} = \theta \\ 0, & \mathbf{x} \neq \theta \end{cases}$.
- $D_w(p, q_\theta) = |\theta|$.

[3]Slide from Stanford CS 236

(Appendix) Wasserstein GAN (WGAN)[4]

- Kantorovich-Rubinstein duality

$$D_w(p, q) = \sup_{\|f\|_L \leq 1} E_{\mathbf{x} \sim p}[f(\mathbf{x})] - E_{\mathbf{x} \sim q}[f(\mathbf{x})]$$

$\|f\|_L \leq 1$ means the Lipschitz constant of $f(\mathbf{x})$ is 1. Technically,

$$\forall \mathbf{x}, \mathbf{y}: \quad |f(\mathbf{x}) - f(\mathbf{y})| \leq \|\mathbf{x} - \mathbf{y}\|_1$$

- Wasserstein GAN with discriminator $D_\phi(\mathbf{x})$ and generator $G_\theta(\mathbf{z})$:

$$\min_\theta \max_\phi E_{\mathbf{x} \sim p_{\text{data}}}[D_\phi(\mathbf{x})] - E_{\mathbf{z} \sim p(\mathbf{z})}[D_\phi(G_\theta(\mathbf{z}))]$$

Lipschitzness of $D_\phi(\mathbf{x})$ is enforced through weight clipping or gradient penalty.

[4]Slide from Stanford CS 236

Issues with Divergences and Density Ratios

- Poor behavior when p^* and q_θ do not have overlapping support.
- Density ratio $\frac{p^*}{q_\theta}$ becomes ∞ or 0.
- $D_{\mathrm{KL}}(p^* \| q_\theta) = \infty$ and $JSD(p^*, q_\theta) = \log 2$.

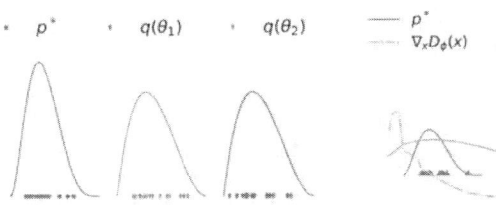

(a) Failure of the KL divergence to distinguish between distributions with non-overlapping support: $D_{\mathrm{KL}}(p^* \| q_{\theta_1}) = D_{\mathrm{KL}}(p^* \| q_{\theta_2}) = \infty$, despite q_{θ_2} being closer to p^* than q_{θ_1}.

(b) The density ratio $\frac{p^*}{q_\theta}$ used by the KL divergence and a smooth estimate given by an MLP, together with the gradient it provides with respect to the input variable.

Figure: Divergences and density ratios.

Contrast with IPMs

- Wasserstein distance and MMD have smoothness requirements.
- Provide useful signal even for distributions with non-overlapping support.

Approximations to f-divergences

- Use a parametric critic D_ϕ.
- Can provide smooth approximations.
- Overcomes non-overlapping support issues.

From Learning Principles to Loss Functions

- Learning principles include class probability estimation, bounds on f-divergences, and IPMs.
- Formulate loss functions for $\boldsymbol{\theta}$ and $\boldsymbol{\phi}$.
- Zero-sum losses: Generator's goal is to minimize the same function the discriminator is maximizing

$$\min \max V(\boldsymbol{\phi}, \boldsymbol{\theta})$$

Or

$$\min \max V(\boldsymbol{\phi}, \boldsymbol{\theta}) = \frac{1}{2}\mathbb{E}_{p^*(\boldsymbol{x})}[\log D_\phi(\boldsymbol{x})] + \frac{1}{2}\mathbb{E}_{q_\theta(\boldsymbol{x})}[\log(1 - D_\phi(\boldsymbol{x}))]$$

Critic and Generator Objectives

- Critic maximizes a quantity to approximate a divergence or distance.
- Model minimizes this approximation.
- Discriminator should distinguish between data and model samples.
- Generator should make its samples indistinguishable according to the discriminator.

Non-zero-sum GANs

- Original GAN is zero-sum.
- Non-zero-sum GAN:

$$\min_{\boldsymbol{\phi}} L_D(\boldsymbol{\phi}, \boldsymbol{\theta}); \quad \min_{\boldsymbol{\theta}} L_G(\boldsymbol{\phi}, \boldsymbol{\theta})$$

where

$$\begin{aligned} L_D(\boldsymbol{\phi}, \boldsymbol{\theta}) &= \mathbb{E}_{p^*(\boldsymbol{x})} g\left(D_{\boldsymbol{\phi}}(\boldsymbol{x})\right) + \mathbb{E}_{q_{\boldsymbol{\theta}}(\boldsymbol{x})} h\left(D_{\boldsymbol{\phi}}(\boldsymbol{x})\right) \\ &= \mathbb{E}_{p^*(\boldsymbol{x})} g\left(D_{\boldsymbol{\phi}}(\boldsymbol{x})\right) + \mathbb{E}_{q(\boldsymbol{z})} h\left(D_{\boldsymbol{\phi}}(G_{\boldsymbol{\theta}}(\boldsymbol{z}))\right) \\ L_G(\boldsymbol{\phi}, \boldsymbol{\theta}) &= \mathbb{E}_{q_{\boldsymbol{\theta}}(\boldsymbol{x})} l\left(D_{\boldsymbol{\phi}}(\boldsymbol{x})\right) \\ &= \mathbb{E}_{q(\boldsymbol{z})} l\left(D_{\boldsymbol{\phi}}(G_{\boldsymbol{\theta}}(\boldsymbol{z}))\right) \end{aligned}$$

Non-zero-sum GANs (Cont'd)

- Functions $g, h, l : \mathbb{R} \to \mathbb{R}$.
  - Original GAN: $g(t) = -\log t$, $h(t) = -\log(1-t)$, $l(t) = \log(1-t)$.
  - Non-saturating loss: $g(t) = -\log t$, $h(t) = -\log(1-t)$, $l(t) = -\log(t)$.
  - Wasserstein: $g(t) = t$, $h(t) = -t$, $l(t) = t$.
  - f-divergences: $g(t) = t$, $h(t) = -f^\dagger(t)$, $l(t) = f^\dagger(t)$.

Nonsaturating Loss

- Nonsaturating loss for better gradient properties early in training L_G:

$$\min_{\theta} \mathbb{E}_{q_\theta(x)} \log -D_\phi(x)$$

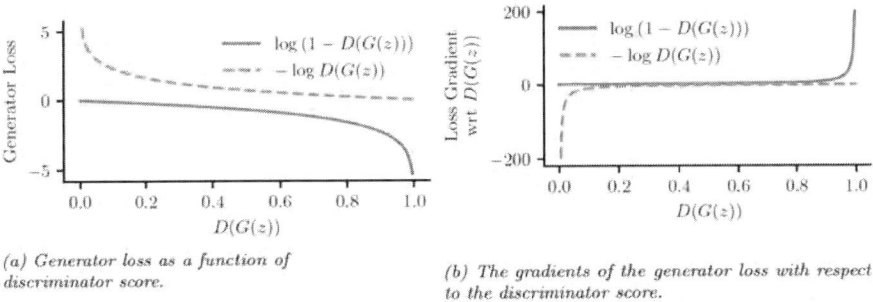

(a) Generator loss as a function of discriminator score.

(b) The gradients of the generator loss with respect to the discriminator score.

Figure: Gradient properties of the non-saturating loss.

Gradient Descent in GANs

- GANs use gradient-based learning for both discriminator and generator.
- Discriminator approximates a distance or divergence $D\left(p^*, q_\theta\right)$.
- Alternating updates between the discriminator and the generator for computational efficiency.

Algorithm for GAN Training

Algorithm General GAN training algorithm with alternating updates

Initialize ϕ, θ
for each training iteration **do**
    for K steps **do**
        Update ϕ using $\nabla_\phi L_D$
    end for
    Update θ using $\nabla_\theta L_G$
end for
Return ϕ, θ

강의노트

1.

2.

3.

에듀컨텐츠·휴피아
Educontents·Huepia

심층생성모델

2024년 12월 10일 초판 1쇄 인쇄
2024년 12월 15일 초판 1쇄 발행

저　　자	임홍기 • 著
발 행 처	도서출판 에듀컨텐츠휴피아
발 행 인	李 相 烈
등록번호	제2017-000042호 (2002년 1월 9일 신고등록)
주　　소	서울 광진구 자양로 28길 98, 동양빌딩
전　　화	(02) 443-6366
팩　　스	(02) 443-6376
e-mail	iknowledge@naver.com
web	http://cafe.naver.com/eduhuepia
만든사람들	기획•김수아 / 책임편집•이진훈 정민경 하지수 박현경 황수정 디자인•유충현 / 영업•이순우
I S B N	978-89-6356-481-4 (93000)
정　　가	18,000원

ⓒ 2024, 임홍기, 도서출판 에듀컨텐츠휴피아

> 이 책은 저작권법에 따라 보호받는 저작물이므로 무단전재와 무단복제를 금지하며, 책 내용의 전부 또는 일부를 이용하려면 반드시 저작권자 및 도서출판 에듀컨텐츠휴피아의 서면 동의를 받아야 합니다.